Shaping an Inclusive Energy Transition

Margot P. C. Weijnen · Zofia Lukszo ·
Samira Farahani
Editors

Shaping an Inclusive Energy Transition

 Springer

Editors
Margot P. C. Weijnen
Faculty of Technology
Policy and Management
Delft University of Technology
Delft, Zuid-Holland, The Netherlands

Zofia Lukszo
Faculty of Technology
Policy and Management
Delft University of Technology
Delft, Zuid-Holland, The Netherlands

Samira Farahani
Faculty of Technology
Policy and Management
Delft University of Technology
Delft, Zuid-Holland, The Netherlands

ISBN 978-3-030-74588-2   ISBN 978-3-030-74586-8   (eBook)
https://doi.org/10.1007/978-3-030-74586-8

This Springer imprint is published by the registered company Springer Nature Switzerland AG
The registered company address is: Gewerbestrasse 11, 6330 Cham, Switzerland

Foreword

Europe aspires to be the first climate-neutral continent by 2050. This goal is necessary. The threat of climate change is existential. It is ambitious. Its achievement requires overhauling what and how we produce and consume. It is feasible if we are all fully committed to it and we make sure no one is left behind.

The transition will be fair or it will not happen. The European Green Deal is our roadmap for this journey. Achieving the climate neutrality goal by 2050 is our common goal but we all start from different points. The distance to travel and the costs the transition entails are not the same for all Member States. Some have a high dependence on fossil fuels and carbon-intensive industries, others are better placed to produce renewable energies.

The same is true for our citizens. All will be actors and subjects of this change but everyone in a different manner. However, if the right mechanisms are put in place at the EU and at the national level, then the European Green Deal will benefit all bringing about new sustainable jobs, a safer and healthier living environment, and a fairer society. At the European level, we are busy putting in place the right policy to finance a fair transition, starting with our Just Transition Mechanism and continuing with a Recovery and Resilience Fund that aims to ensure we will be able to build back better after the COVID-19 crisis.

The decarbonisation of our energy system is a necessary step towards climate neutrality. Technological innovation and change will be at the heart of a transition that will see a massive shift towards renewable energy resources and climate-neutral energy services.

The magnitude of the necessary transformation and the challenges this raises can hardly be underestimated. Systemic change will require concerted action along the value chains of electricity, heating, cooling and mobility services, from production to energy transport, storage and use. Each stage will involve massive investments. For this to happen, the right political and policy frameworks are needed to dispel the technological and social uncertainties any such transformation would imply.

The transition is further complicated by the increasing interdependencies within the multi-carrier energy system, and between the energy, transport and digital infrastructure systems. In this complex setting, concerted action is not a given, even if all parties embrace the Green Deal. Governance becomes the essential issue and it is here that the proposed European Climate Law shows its fundamental value. By proposing a legally binding target of net zero greenhouse gas emissions by 2050 and by defining intermediate goals for emission reductions on the trajectory between 2030 and 2050, the European Climate Law will fix the long-term destination of travel, allow progress to be measured and reduce uncertainty for public authorities, businesses and citizens.

Our first intermediary enhanced target for greenhouse gas emission reductions in 2030 of at least 55% on a net basis compared to 1990 has been agreed. To reach it, the European Commission is reviewing and will propose revisions of all policy instruments needed.

This book offers public policy and engineering professionals guidance in navigating the complexity of the energy system and it provides practical perspectives for bringing the goal of climate neutrality into being. Most of all, it also takes into account the social dimension of the energy transition and re-emphasises social inclusiveness as a key value for the very success of our pursuit of a climate neutral energy system.

We are the first generation to feel the impact of climate change and we are the last that can do something about it. I hope the chapters in this book provide you with ample inspiration on what to do.

Brussels, Belgium   Frans Timmermans
March 2021   Vice-President of the European Commission

Preface

The idea for this book dates back to September 2018, when I organised a symposium to mark the occasion of the inaugural address of my dear colleague Zofia Lukszo as a full professor of Smart Energy Systems at the Delft University of Technology. In joint recognition that making energy systems 'smarter' is not a goal in itself, we realised that the higher goal of our research efforts is to help ensure the availability of sustainable, reliable and affordable energy services for all members of society. As energy is the lifeblood of society and the economy, access to energy services is an essential condition not only for businesses to be productive, but also for the wellbeing and welfare of each citizen. Without energy services, people are largely deprived of possibilities to contribute meaningfully to modern society and the economy. Hence, the symposium and this book started from the recognition that the challenge of the energy transition entails more than decarbonising the energy system. While this is a formidable challenge in itself, the real challenge is to ensure that nobody is left behind in the energy transition. In the words of Ursula von der Leyen, upon the presentation of the European Green Deal, it is all about *a transition that is just and socially fair*. We wholeheartedly support this view of the energy transition. To leave no doubt about the normative stance we take in this book, we adopted the title: Shaping an Inclusive Energy Transition.

The multi-scale and multi-dimensional complexity of the energy system implies that the transition is not one clear trajectory. Rather, it is a web of trajectories, for different energy carriers, with many actors intervening simultaneously, at different scales, with different time horizons. The messy reality of the energy system defies any claims of rigour and comprehensiveness in shaping the energy transition at the global energy system level. We see the energy system as a system of intricately interwoven, constantly interacting supply chains providing us with electricity, transport and heating fuels. It is a system deeply embedded in the spatial and economic structure, in the built environment and in our social routines. It involves multiple markets and physical networks extending across national borders, thus involving multiple national jurisdictions. It cuts across the public and the private sector in different ways in different countries. It is a system populated with a multitude of actors: (supra)national, regional and local governments, producers of energy resources, energy suppliers, network providers, energy service providers, technology providers, market operators, network regulators and so forth and, last but not least, a wide range of end-users in all sectors of the economy and society. Each one of these actors upholds a different set of values, has different interests in the energy transition and different means at his disposal to steer the system on the transition path. Moreover, some actors play multiple roles, for example, as citizens and as consumers.

In this view, the energy system is a complex adaptive system, a system that is constantly evolving as a result of the operational and strategic decisions made by the actors. Many of these decisions have short-term local effects; some have long term and supra-local effects; decisions by one actor influence the decisions of other actors; some decisions may lead to new actors entering the system or result in established actors acquiring new roles. Together, the decision making of autonomous actors in the multi-actor system influences the web of constantly changing interactions in the system, including a variety of feedback loops across system levels and time scales. Only from observing the emergent overall behaviour of the system can we see if the system performs and evolves according to our expectations and priorities.

The complex adaptive nature of the energy system defies notions of 'shaping' the energy transition. The transition is not malleable. It cannot be designed as in engineering design. We do, however, use the term shaping to denote that a concerted effort is needed, and that we have a responsibility to shape our interventions in the energy system with respect for the values at stake, including social and ethical values of equity, fairness and justice. Interventions can be of a technological nature, they may come in the form of legislation and regulations, they may be financial (dis)incentives to encourage certain investments, they may involve land reservations, they may be soft incentives to nudge our behaviour towards reducing our energy demand, and so forth. The list of possible intervention options is sheer endless. The challenge is to design the set of interventions smartly in such a way as to make them re-enforce each other for maximum effect in directing and accelerating the transition. If and how and to what extent interventions can be aligned depends on the governance of the energy system. The governance of the energy system is not easily delineated in terms of actors, rules, norms and decision-making processes, as energy is everywhere. While national governments typically have a ministry for energy, many decisions made in ministries of transport, industry, agriculture, environment, et cetera can also heavily impact the energy system. In shaping the energy transition, one of the challenges is to align these silos of decision making. Even within the traditional boundaries of energy system governance, we often see a siloed structure that makes decision makers blind to the interactions between electricity, transport fuels, natural gas and other heating fuels within the multi-carrier energy system.

Another complication is that the energy transition has many different starting points, as each country has its own mix of energy resources, its own legacy infrastructure and its own economic structure. The challenge is furthermore complicated by cultural differences. These are reflected in a different governance culture and structure, and in different traditions and behavioural patterns with regard to energy use.

While we cannot and do not pretend that this book provides a comprehensive and definitive answer as to how to shape an inclusive energy transition, we do hope it is useful to inspire strategic decision makers engaged in the energy transition, especially public policy makers, to increase their awareness of potential unintended consequences of their decisions, and make them reflect on their own values as much as on the social values at stake. It provides illustrative examples of how particular technology and governance choices may help to push the energy transition, how such choices may affect people's lives, and how citizens may be engaged in the energy transition.

In this book, various authors explore the solution space for a future sustainable and inclusive energy system. Their disciplinary angles vary widely, ranging from the engineering sciences to public management, law, philosophy and ethnographic studies. In their respective contributions they adhere to the concepts and terminology that are prevailing in their discipline and recognised by the practitioners in their area. We refrained from an ontological analysis of energy infrastructure and the energy transition, as an ontology of key concepts would mainly be of academic interest. For public policy makers and other practitioners engaged in shaping the energy transition, which we see as the target audience for this book, the value of such an ontology is, at best, limited.

In the first part of this book, we show how the transition is grounded in the infrastructure legacy, in which the norms, values and ambitions of our forefathers are expressed. We describe how the changing values and priorities of today's society cause friction with the established infrastructure and governance structure, and provide examples of interventions by which such friction may be overcome. Energy justice is one of the areas where friction is evident, which must be accounted for in designing policy interventions. Moreover, it is brought to the fore that in any intervention we intend to design, whether technological, political or policy-wise, existential questions arise about our own individual value systems as the hidden dimension in the decisions and actions we take.

In the second part, we delve deeper into two promising directions of technological innovation: the emergence of smart grids and the development of hydrogen as a future energy carrier substituting for natural gas. Both developments have far reaching consequences for the design and operation of the future energy system, and both have huge potential for solving one of the most prominent problems of a future energy system based on renewable energy resources, which is: how to deal with the natural weather dependent and geospatial variability of wind and solar power supply? Even more intriguingly, both developments contribute to an increasingly intimate interconnection between energy infrastructures in the multi-carrier energy system, and between the energy system, the transport system and the IT and telecommunications system.

In the third part, the focus is on the institutional dimension of the energy transition. It starts with the condition of societal appreciation as a condition for the acceptance of new policies and policy instruments, as illustrated on the basis of an EU wide study. It is argued that any change in the provision of energy services needs a strong legal framework, so as to protect consumers, especially in the case of captive users connected to heat distribution networks. A case study of Sofia, Bulgaria, illustrates how heat provision is at the heart of the energy-water-food nexus in the city, and how a nexus approach may help to solve the huge problem of energy poverty for a large group of vulnerable users.

In the last part of this book, we address the challenge of enabling public participation in shaping an inclusive energy transition. After all, if the energy transition is to deliver on its promise, it will benefit all of us. In a truly inclusive energy transition, it is not some higher authority deciding how we will benefit, without consulting us. As citizens and energy users, we should be enabled to bring in our own views, values and opinions to weigh in the decision making, especially where this affects our own households, neighbourhoods and living environment. Shaping an inclusive energy transition is not only about the inclusiveness of the future energy system, but also about the process of engagement and decision making. Hence, in this process, no one should be left behind.

Last but not least, in the final chapter, a comprehensive engineering systems approach is proposed to acknowledge the interactions between technology and institutions in shaping the future energy system. As such it helps engineers and policy makers to engage in joint pursuit of the energy transition, and more precisely, to engage in constructive debate about the shaping of interventions in the energy system. This interaction is crucial, since many technological interventions cannot be brought into being and will not come to fruition if appropriate institutions are lacking, and the other way around: the institutional setting is generally not technology-neutral. Policy makers must be aware of technology bias in the institutional design of policy interventions.

I would like to thank all contributing authors for their efforts in bringing this book into being. All of you have generously shared your vast knowledge for the sake of this book and the corresponding massive open online course on Inclusive Energy Systems. I would also like to thank the students in this course for their valuable questions and comments. You challenge us and keep us focused.

Finally, I would like to acknowledge the generous support of Next Generation Infrastructures, the Dutch national knowledge platform of infrastructure providers, in funding this book and the workshop in May 2019, in which the authors fleshed out the idea for this book. Next Generation Infrastructures advocates a cross-sector system-of-systems approach to infrastructure development, as a logical consequence of the more and more intricate interactions and interdependencies between infrastructures, which are, however, not yet reflected in the current governance structures. Moreover, Next Generation Infrastructures calls for a new public debate on infrastructure, in which the social value created by infrastructure takes centre stage. Infrastructure, including energy infrastructure, is not a goal in itself, but only a means to enable the society we want to be. This message resonates with the message of this book, in which we embrace energy infrastructure and the energy transition as a means to bring a more inclusive society into being.

February 2021

Margot P. C. Weijnen
Professor of Process and Energy Systems
Engineering, Faculty of Technology
Policy and Management, Delft University
of Technology
Delft, The Netherlands

Contents

Connecting Technology and Society

About the Editors

Prof. dr. ir. Margot P. C. Weijnen is a full professor of Process and Energy Systems Engineering at the Delft University of Technology, since 1995. She is the founding and scientific director of Next Generation Infrastructures, the Dutch national knowledge platform of infrastructure providers, since 2001. She was a member of the Governing Board of the IEEE System, Man and Cybernetics Society and established the IEEE SMC Technical Committee for Infrastructure Systems & Services. She has been engaged in numerous advisory capacities to the Dutch government and the European Commission as a member of e.g., the EC Advisory Group on Energy, the Netherlands Council for Science and Technology Policy, the Dutch General Energy Council, and the Netherlands Scientific Council for Government Policy. She chairs the Advisory Board of the Royal Netherlands Aerospace Centre and is a member of the Supervisory Board of AkzoNobel Netherlands BV and Shell Netherlands BV. Since March 2020, she sits on the Executive Board of the Dutch National Research Council, where she chairs the Executive Board of the Engineering and Applied Sciences domain.

Prof. dr. ir. Zofia Lukszo is a full professor of Smart Energy Systems. She heads the Energy and Industry Group at the Faculty of Technology, Policy and Management at Delft University of Technology. The group is involved in research and education in the field of energy infrastructure systems and focuses on the innovative development of technological 'software' i.e. structured methods and tools to support the energy transition. An important theme in the research programme concerns modelling and simulation of large-scale infrastructure networks as socio-technical systems using an agent-based approach. Her research concentrates on a wide range of problems in the way complex energy systems are functioning and can be (re-)shaped for the sustainable future. Shaping smart grids, which require additional flexibility from the users in interaction with dispatchable power plants, storage and interconnections, is one of the main pillars in her research. Zofia is a Steering Board member of PowerWeb Institute—an interfaculty research programme on smart and integrated energy systems, and a Supervisory Board member of the Green Village—a living lab connected to the TU Delft. As a senior IEEE Member she contributed to many international workshops and conferences. She has over 150 refereed scientific publications and book contributions.

Dr. ir. Samira Farahani is an expert in energy system design and analysis with more than 10 years of experience. She has obtained her Ph.D. degree in Systems and Control from Delft University of Technology and since then, she has been working on several projects in the area of smart energy systems focusing on data-driven analysis and optimal operation. At the time of preparation of this book, she was appointed as a senior researcher at the Faculty of Technology, Policy and Management at Delft University of Technology. Prior to this, she was business strategy manager at the consulting company Accenture. In the past years, she has also worked at prestigious institutes such as TU Delft, Caltech, and Max-Planck Institute, and has collaborated with several companies such as Eneco, Shell, Stedin, Gasunie, etc. in different projects. Her main area of expertise is in design and analysis of (smart) integrated energy and mobility systems, with emphasis on the role of hydrogen as a new energy carrier in the low-carbon energy transition. Samira is joining the non-profit organization WaterstofNet, as project manager, in February 2021 to continue supporting the hydrogen development activities in Benelux.

Introduction

Margot Weijnen, Zofia Lukszo, and Samira Farahani

A Wake-Up Call to Inequality

The energy transition is an unprecedented challenge for the world. It is unprecedented in its global ambition and in its complexity. In the United Nations Agenda for Sustainable Development (UN, 2015a) it is formulated as 'Achieving Sustainable Energy for All'. While it is only one of the seventeen Sustainable Development Goals to be accomplished by 2030, the agenda emphasises that the Sustainable Development Goals are strongly interrelated. The goal of achieving sustainable energy for all is crucial indeed in achieving many of the other development goals. Health, food security, gender equality, education, economic development and other sustainable development goals critically depend on access to clean, affordable and reliable energy services.

Sustainable energy for all implies access to energy services for all world citizens. The International Energy Agency (IEA) defines energy access as "a household having reliable and affordable access to both clean cooking facilities and to electricity, which is enough to supply a basic bundle of energy services initially, and then an increasing level of electricity over time to reach the regional average" (IEA, 2020a). For electricity, a basic bundle of energy services means, at a minimum, several lightbulbs, task lighting (such as a flashlight), phone charging and a radio. This minimum service level covers only bare necessities. Most world citizens need a lot more electricity to meet their energy service demands. Hence, the IEA added the notion of regional average to indicate that the minimum energy service level that defines energy access also depends on where and how people live. The minimum

M. Weijnen (✉) · Z. Lukszo · S. Farahani
Faculty of Technology, Policy and Management, Delft University of Technology, Jaffalaan 5, 2628 BX Delft, The Netherlands
e-mail: m.p.c.weijnen@tudelft.nl

Z. Lukszo
e-mail: z.lukszo@tudelft.nl

© The Author(s) 2021
M. P. C. Weijnen et al. (eds.), *Shaping an Inclusive Energy Transition*,
https://doi.org/10.1007/978-3-030-74586-8_1

energy requirements may differ for city residents and rural dwellers, they depend on the state of economic development of the country or region and they are defined by geographic conditions. For instance, people living in a cold climate know that their life may literally depend on adequate heating services during winter time.

Between now and 2030, there is still a long way to go. Since 2000, the number of people without access to electricity has declined from 1.7 billion to 1.1 billion in 2016 and to 770 million in 2019 (IEA, 2020b). This remarkable progress was mainly achieved by grid expansion and new power generation capacity on the basis of fossil fuels. However, although electricity access is improving and the contribution of renewable energy sources is rapidly increasing, it still is a formidable challenge to secure access to electricity for the entire world population by 2030. This is not only about securing a bundle of essential energy services for households, but also about the demand for electricity in industry, the agricultural sector, and the service sectors of the economy. Even with the most energy efficient technologies currently available, industry and services need grid-based access to be productive.

Even more urgent than access to electricity, is access to affordable and cleaner fuels for cooking and heating. In sub-Saharan Africa and remote regions of Asia, many countries still rely on solid biomass, which in practice often implies that women and children spend hours per day to collect firewood. This reduces their opportunities for education and economic activities. Moreover, the indoor use of firewood and other poor-quality fuels for cooking or heating has detrimental health effects. Worldwide, close to 4 million premature deaths per year are attributed to the indoor use of polluting cooking and heating fuels (World Health Organization, 2018).

While installed capacity for power generation from renewable energy sources is rapidly increasing, the energy transition challenge involves more than decarbonising the electricity system. It also affects the provision of fuels for cooking, heating and transport services. Electricity is by far not the largest part of the final energy consumption. For the world, the share of electricity in the total final energy consumption amounted to 19.3% in 2018 (IEA, 2020c). In the EU 27, in 2018, the share of electricity in the total final energy consumption amounted to 20.8%, while petroleum products represented a share of 40.8%, and natural gas a share of 20.8% (European Commission, 2019). In the Netherlands, the share of natural gas in the 2018 final energy consumption amounted to 34.1%, far more than anywhere else in the EU 27. For the share of solid fuels (mostly coal) in energy end consumption also huge differences are observed between EU Member States, with a mere 0.3% in the Netherlands and 15.6% in Poland. The huge share of petroleum products in the energy consumption mix is to a large extent explained by their prominent role as transport fuels and industrial heating fuels: the industry sector and the transport sector represent 32% and 28%, respectively, of the overall EU 27 final energy consumption, while the share of households is only 24% (European Commission, 2019). These figures help us to understand the enormity of the decarbonisation challenge. While the EU is on track in view of its renewable energy targets for 2020, as shown in Fig. 1, a considerable gap needs to be bridged to reach the 32% goal for renewables in 2030, besides a 40% reduction of greenhouse gas emissions. The newly proposed European Climate Law

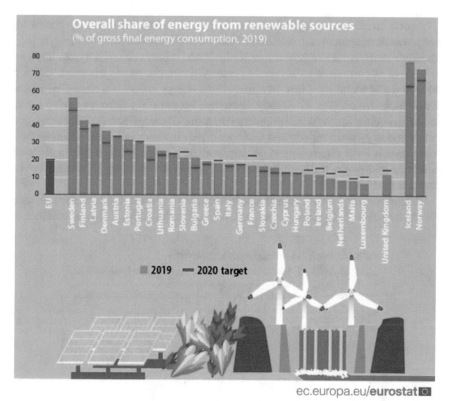

Fig. 1 Share of energy from renewable sources, 2019 (% of gross final energy consumption). *Source* Eurostat (2020)

is even raising the ambition level of greenhouse gas emissions reductions from 49 to 55% in 2030 (compared to 1990 emission levels).

The annual monitoring reports of the World Economic Forum (WEF) on how countries around the globe are "Fostering Effective Energy Transition" reveal large differences between countries in energy transition readiness and progress towards a sustainable, affordable, secure and reliable energy system (World Economic Forum, 2020). Also, within the group of the world's largest energy consumers, the WEF finds big differences in energy transition performance. While India and China show a steady improvement, the performance indicators for the United States and Brazil are declining. This is a reason for concern, as a successful and timely global scale energy transition hinges on the performance of the largest energy consumers and their willingness to lead the transition. A glimmer of hope may be found in the observations that the gap between the top energy transition performers and the rest is steadily decreasing, that the energy intensity of GDP is generally lower in rich countries and that the energy intensity of GDP on a global scale has declined over time (Stern, 2018; World Economic Forum, 2020).

Access to affordable and reliable energy services is an essential condition for economic development and wellbeing. This goes for national economies and for individual citizens. A lack of access to affordable and reliable energy services deprives citizens of opportunities to engage in personal development and economic activity, even in many social activities. In other words: energy access is a crucial condition for an inclusive society. As already discussed, we are still far from catering for an inclusive society for the world population in terms of energy access. The correlation between energy consumption and GDP per capita is shown in Fig. 2, which furthermore reveals large differences in per capita energy consumption between different countries and world regions, and therewith in development opportunities for the population (IEA, 2017).

Energy consumption is connected to income, but the link is loose and variable over time; many countries reach a point where they can reduce energy use and continue to

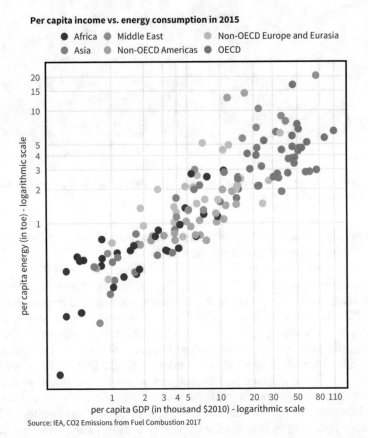

Fig. 2 Correlation of energy consumption and GDP per person; the graph shows per capita energy consumption (in tonnes of oil equivalent) versus per capita GDP (in thousand $2010). All values refer to the year 2015 (IEA, 2017)

grow economically. There is a clear link between per capita energy demand and per capita gross domestic product (GDP) as can be seen in Fig. 2: there are no developing countries that use a lot of energy per capita, and there are no developed countries that consume as little energy as the developing ones. But beyond that, the relationship is hard to define. Countries with similar incomes often consume two, three, four, or five times more energy between them (note the scale is logarithmic). It is similarly common for countries with vastly different incomes to use the same amount of energy. GDP matters, in other words, but so does climate, economic structure, whether energy is taxed or subsidized, technology, policy, and so on (Tsafos, 2018).

Today's situation of almost 800 million people without access to electricity and many more without access to clean fuels for heating and cooking can thus be seen as a case of energy injustice, and it can be considered a moral imperative to remedy this injustice. In practice, lack of access to energy services can be attributed to different causes requiring different solutions. It is not only a problem of poverty; in many developing countries, also people who can afford electricity services are not given physical access to the supply. In such cases, structural issues are often at play, such as inadequate policy and lack of infrastructure. In developing economies, we often see a pattern where electricity infrastructure is provided in the cities, while the less densely populated rural areas are lagging behind. In developed countries, issues of energy access are generally not due to a lack of physical access. Rather, they pertain to the affordability of energy services. That is why the term energy poverty is often used, which is an indicator related to the pressure of energy costs on the disposable household income. Most countries use a threshold of 10% of the household income to indicate energy poverty. Energy poverty is a widespread problem in developing and in developed economies alike. Even in OECD[1]-economies, around 200 million people suffer from energy poverty according to IEA estimates (OECD/IEA, 2017), that is more than 15% of the total population. For the energy poor in OECD countries, energy poverty is often related to the poor quality of their homes, which implies high heating costs to reach a minimum standard of comfort in cold times, if it can be reached at all.

As the world stands, the provision of reliable and affordable energy services is not yet a given for all nations and world citizens. Where such services—and other essential infrastructure services—are lacking, socioeconomic development is impeded. Stark differences in development opportunities by region may contribute to massive migration flows with potentially destabilising consequences. Even within developed economies, large differences in personal development opportunities between groups of the population are a recipe for instability. Access to energy services is one of the conditions, like access to other infrastructure services, for citizens to engage in personal development and economic activity, in other words, to contribute to society and create an inclusive society. That is the compelling reason to act in pursuit of the UN Sustainable Development Agenda at the global and at the local level.

---

[1]Organisation for Economic Co-operation and Development is an intergovernmental economic organisation with 37 member countries, founded in 1961 to stimulate economic progress and world trade. www.oecd.org.

Sustainable Energy

The energy transition is about more than providing access to energy services to all world citizens. It is also about the sustainability of energy services. It is about preserving the quality of the living environment and the regenerative capacity of the natural system. In the previous century, policy ambitions were predominantly focused on reducing local and regional impacts of the energy system on public health, safety and the natural environment, such as caused by emissions of particulate matter, SO_x and NO_x emissions to the atmosphere, solid waste and emissions to surface waters. Since then, policy focus has shifted to combating climate change, as science has provided compelling evidence that the progressive large-scale exploitation of fossil fuels since the industrial revolution is to blame for climate change unfolding at an increasingly rapid pace. Due to the abundant combustion of fossil fuels, the carbon dioxide concentration in the air we breathe has increased from 300 ppm, before the industrial revolution, to more than 400 ppm today. As a consequence, the natural greenhouse effect of the earth's atmosphere is increasing to levels where the future liveability of our planet for human beings is at stake. A glimmer of hope that the Anthropocene may not be a prelude to the end of human life on our planet, may be found in the truly global scale of climate policy action, as embodied in e.g., the Kyoto protocol and the more recent Paris agreement. Although the effectiveness of global climate policy may be questioned, it is one of the very few examples, like the establishment of the United Nations itself, where so many countries worldwide have agreed to overcome purely national interests and join in a concerted effort for a global cause (UN, 2015b).

Today, as we have come to realize that CO_2 emissions resulting from the combustion of fossil fuels are to blame for the threat of climate change at a global level, policy focus has shifted to decarbonising the current energy system and accomplishing a transition towards a system based on renewable energy resources. Both are formidable challenges which can greatly be facilitated by a structural reduction of energy demand in all sectors of the economy and society. Since energy use is deeply embedded in the routines and structure of all sectors of the economy, this is a difficult challenge in itself. A structural reduction of energy demand that goes beyond energy efficiency measures not only requires massive investment, but also the adoption of new design principles in systems, buildings and processes, and behavioural change in many of our daily routines. Decarbonisation of the energy system can only be achieved by carbon capture and sequestration as long as fossil fuel resources have not yet been replaced by renewable energy resources. Meanwhile, even though the installed renewable energy capacity is growing at an impressive pace, the supply of renewable energy services can hardly keep up with the increasing global energy demand.

Another driver for the transition towards renewable energy resources is the wish to reduce our exposure to the geopolitical risks and vulnerabilities that are inherent to the skewed geographical distribution of fossil fuel resources (Bartuška et al., 2019; Correljé & van der Linde, 2006). Especially oil reserves are controlled by a limited number of countries, most of them represented in OPEC, which holds the potential of market power abuse. Another vulnerability is posed by the oil transportation routes. Oil carriers using shipping routes through narrow straits and oil pipelines crossing politically instable areas are potential targets for terrorist attacks. Disruption of oil supply chains can potentially disrupt oil dependent economies. This dependence is especially strong in the transportation sector, as the high energy density transportation fuels currently used in aviation, shipping and heavy road transport are all derived from oil. Another sector characterized by a high dependency on oil (and other fossil fuels) is the process industry, as many base chemicals (many of which are derived from oil) and base metals require high temperature processes.

In hindsight, it is too easy to judge the use of fossil fuels, even their current use, as irresponsible now that we have become aware of the potentially dire consequences. Thanks to the industrial revolution, large parts of the global population have been lifted out of poverty. It is largely thanks to fossil fuels that reliable energy services have come within reach of most of the global population, in terms of physical access and affordability. It is a challenge to stage the transition to a truly sustainable climate neutral energy system in such a way so as not to thwart the development opportunities for countries and individuals which have thus far been deprived of such opportunities.

Transition or Disruption?

In using the term energy transition, it is generally assumed that the change process from a fossil fuel-based energy system to a sustainable energy system can be organized and managed in an orderly fashion. There seems to be an implicit assumption that the transition can, at least to a large extent, be designed as a sequence or constellation of technological innovations and behavioural changes, which can be forged by appropriate policy measures, e.g., incentives and regulations. The idea of transition does not deny the need for radical change, but presupposes that social and economic disruption can be avoided.

The assumption of a manageable transition may need to be re-examined in the light of the crises faced by the world in the twenty-first century, such as the financial and economic crises of the first decade and the social and economic crisis caused by the Covid-19 pandemic today. As it turns out, the pandemic is exacerbating the socioeconomic inequalities in the world, between nations, and within nations between segments of the population.

As it grapples with the unprecedented health emergency triggered by the Covid-19 pandemic, the world is experiencing its worst economic shock since the 1930s. This is having a severe impact on employment and investment across all parts of the economy, including the energy sector. Due to long periods of quarantine in many countries all around the world, the energy consumption and production patterns have also changed. An analysis by Wärtsilä Energy Transition Lab for the period March to April 2020, reveals that, compared to the same period in 2019, the coal fired power generation in Europe has been reduced by 29%, the CO_2 intensity was reduced by

20%, the share of renewable generation increased by 8%, reaching 46% in the total energy mix, and finally the energy demand was reduced by 10% compared to the same period in 2019 (Wärtsilä Energy Transition Lab, 2020). At the same time, the world experienced a drastic crude-oil price reduction in April 2020, due to political disagreements between OPEC countries as well as the demand reduction caused by the Covid-19 pandemic.

Governments have taken the lead in providing urgent financial and economic relief to prevent the crisis from spiralling further downward [cf. (BBC News, 2020; German Federal Ministry of Finance, 2020; Government of the Netherlands, 2020; Magazine, 2020]. "Today, attention is increasingly focusing on how to bring about an economic recovery that repairs the damage inflicted by the crisis while putting the world on a stronger footing for the future," as stated by the IEA Executive Director (IEA, 2020d). To assist a quicker recovery, IEA has published a plan in June 2020 (IEA, 2020d). The recovery plan suggested by IEA is focused on three main goals: boosting economic growth, creating jobs and improving future sustainability and resilience of the energy system. As shown in Table 1, they have analysed a range of energy-related measures, which countries could adopt in their recovery plans.

Furthermore, to help repair the economic and social damage caused by the coronavirus pandemic, the European Commission, the European Parliament and EU leaders have agreed on a recovery plan that will lead the way out of the crisis and lay the foundations for a modern and more sustainable Europe (European Commission, 2020). Based on the achieved agreement, more than 50% of the supporting fund goes to innovation and research, climate and digital transition, and health facilities. It is noticeable that 30% of the EU funds is dedicated to climate change mitigation activities, the highest share ever of the European budget. Moreover, in the NextGenerationEU, which is a €750 billion temporary recovery instrument to help repair the immediate economic and social damage brought about by the pandemic, one of the focus points is supporting the Member States with investments and reforms to reach their climate policy goals (European Commission, 2020).

The pandemic has both negative and positive effects on the energy transition. Its negative impact is related to a massive increase in uncertainties which resulted in many investment projects being put on hold. The potential long-term consequence of a substantial delay in investments is utter failure to fulfil the climate targets both worldwide and at the country-level. That is why the recovery plans of many countries entail the provision by governments of large amounts of investment capital to support new and ongoing energy transition projects (see for instance (BBC News, 2020; German Federal Ministry of Finance, 2020; Government of the Netherlands, 2020; Magazine, 2020)). On the other side, in the perspective of the energy transition, the pandemic also brings positive effects in terms of reducing energy demand, coal consumption and CO_2 emissions. The challenge now is to maintain these downward trends in the post-Covid-19 era, and to continue working on achieving the planned targets. Moreover, as suggested by the IEA recovery plan, the energy transition will create new employment opportunities in the field of energy efficiency, renewable

Table 1 IEA energy sector measures to be considered in the recovery plan by the governments (IEA, 2020d)

Sector	Measure
Electricity	• Expand and modernise grids • Accelerate the growth of wind and solar PV • Maintain the role of hydro and nuclear power • Manage gas- and coal-fired power generation
Transport	• New vehicles • Expand high-speed rail networks • Improve urban infrastructure
Buildings	• Retrofit existing buildings and more efficient new constructions • More efficient and connected household appliances • Improve access to clean cooking
Industry	• Improve energy efficiency and increase electrification • Expand waste and material recycling
Fuels	• Reduce methane emission from oil and gas operations • Reform fossil fuel subsidies • Support and expand the use of biofuels
Strategic opportunities in technology innovation	• Hydrogen technologies • Batteries • Small modular nuclear reactors • Carbon capture, utilisation and storage

energy production, infrastructure and services, and in research, development and innovation.

The Way Forward

Access to energy services, and the availability and affordability of energy services, are crucial conditions for social and economic development, for individual citizens and society as a whole. At the same time, these crucial services must be acceptable for society, which is to say that they must comply with the values and priorities of society. Over time, these values and preferences are changing. For a long time, acceptability of energy services was more or less synonymous with health, safety and environmental issues that could largely be solved by technological means. Today, societal priorities have come to include the combat of climate change, and the meaning of acceptability has come to include issues of equity, fairness and justice. At the most fundamental level, it is this change in societal values and priorities which is steering the energy transition.

In the pursuit of the energy transition it is a major challenge to connect technological change to the social values at stake. The feasibility of the energy transition hinges on public acceptance of the massive investments required as well as the behavioural changes called for in the way we use energy. Energy services are necessary to enable economic activity and to support us in our daily routines, in virtually everything we do, not only as economic agents, but also as members of social communities to which we contribute. This awareness is crucial for all involved in the energy transition to acknowledge its social dimension, beyond the technological challenge. Rather than deepening existing social divides or causing new rifts, we all have a moral obligation to shape the energy transition in such a way as to support a more inclusive society.

References

Bartuška, V., Lang, P., & Nosko, A. (2019). The geopolitics of energy security in Europe. In: *New perspectives on shared security: NATO's next 70 years*. November 28, 2019.

BBC News. (2020). *France in huge coronavirus recovery plan focusing on green energy*. https://www.bbc.com/news/business-54009642

Correljé, A., & van der Linde, C. (2006). Energy supply security and geopolitics: a European perspective. *Energy Policy* (34)5:532–543.

European Commission. (2019). *Renewable energy statistics*. https://ec.europa.eu/eurostat/statistics-explained/index.php/Renewable_energy_statistics

European Commission. (2020). *Recovery plan for Europe*. https://ec.europa.eu/info/strategy/recovery-plan-europe_en

Eurostat. (2020). *Share of the renewable energy in the EU*, Dec 2020. https://ec.europa.eu/eurostat/web/products-eurostat-news/-/ddn-20201218-1

German Federal Ministry of Finance. (2020). *Emerging from the crisis with full strength*. https://www.bundesfinanzministerium.de/Content/EN/Standardartikel/Topics/Public-Finances/Articles/2020-06-04-fiscal-package.html

Government of the Netherlands. (2020). *Recovery plans for 2021*. https://www.government.nl/government/the-government-s-plans-for-2021

IEA. (2017). *CO_2 emissions from fuel combustion*. All rights reserved.

IEA. (2020a). *Defining energy access: 2020 methodology*. International Energy Agency. https://www.iea.org/articles/defining-energy-access-2020-methodology.

IEA. (2020b). *SDG7: data and projections*. International Energy Agency. https://www.iea.org/reports/sdg7-data-and-projections.

IEA. (2020c). *Key world energy statistics 2020*. International Energy Agency. https://www.iea.org/reports/key-world-energy-statistics-2020/final-consumption.

IEA. (2020d). *Sustainable recovery*. World Energy Outlook Special Report. International Energy Agency, June 2020. https://www.iea.org/reports/sustainable-recovery.

Magazine, P. V. (2020). *Australian states target green recovery*. https://www.pv-magazine.com/2020/09/14/covid-19-weekly-round-up-australian-states-target-green-recovery-as-france-devotes-e30bn-to-energy-transition/.

OECD/IEA. (2017). *Energy access outlook 2017—from poverty to prosperity. World Energy Outlook special report*. International Energy Agency. pp 24–25, Fig 1.3.

Stern, D. I. (2018). Energy-GDP relationship. In: Macmillan Publishers Ltd (eds) *The new palgrave dictionary of economics*. Palgrave Macmillan. https://doi.org/10.1057/978-1-349-95189-5_3015

Tsafos, N. (2018). *Must the energy transition be slow? Not Necessarily*. Center for Strategic & International Studies, CSIS Briefs, September 14, 2018. https://www.csis.org/analysis/must-energy-transition-be-slow-not-necessarily.

UN. (2015a). *Sustainable Development Agenda*. United Nations. https://www.un.org/sustainabledevelopment/development-agenda/.

UN. (2015b). *Adoption of the Paris agreement framework convention on climate change*. United Nations. https://unfccc.int/resource/docs/2015/cop21/eng/l09r01.pdf.

World Economic Forum (WEF). (2020). *Fostering effective energy transition. Country transitions and benchmarking*. https://www.weforum.org/projects/fostering-effective-energy-transition.

Wärtsilä Energy Transition Lab. (2020). *Press release*, April 17, 2020.

World Health Organization (WHO). (2018). *Household air pollution and health*. https://www.who.int/news-room/fact-sheets/detail/household-air-pollution-and-health.

Infrastructure and Values

Rethinking Infrastructure as the Fabric of a Changing Society

With a Focus on the Energy System

Margot Weijnen and Aad Correljé

Abstract In this chapter, we explore the nature of infrastructure, how it is appreciated by society, how this appreciation has changed over the lifetime of the infrastructure, and how infrastructure development and performance are influenced by the governance structures in place. While the focus in this chapter is on energy infrastructure, ample illustrative material is also provided from other infrastructure sectors. We examine the trends towards technological and administrative decentralisation and towards digitalisation of infrastructure (service) provision. These trends enable formerly passive consumers to adopt new roles as providers of energy, data and transport services, and result in strongly increasing cross-sector interdependencies, especially between energy, transport and digital infrastructure. These interdependencies, however, are not reflected in the siloed governance structure of these domains, which hinders the energy transition. Furthermore, we diagnose a mismatch between, on the one hand, the focus of energy infrastructure governance on cost-effectiveness—with a view to low-cost service provision—and, on the other hand, the role of infrastructure in upholding and creating social value in terms of equity, fairness and social justice. Since the energy market liberalisation, the fundamental role of infrastructure as the *fabric of society* appears to be a blind spot in reflections on infrastructure and largely unexplored territory in current infrastructure policy and governance. If not remedied, this blind spot may exacerbate existing inequalities between energy consumers and create new divides in society, as is illustrated by current developments in the Netherlands with respect to sustainable heat provision. We advocate a richer value orientation in energy infrastructure governance and infrastructure governance at large, which goes beyond the current focus on efficiency and economic value, in recognition of changing societal values and priorities and, most of all, to fulfil the potential of infrastructure in creating an inclusive society.

M. Weijnen (✉) · A. Correljé
Delft University of Technology, Delft, The Netherlands
e-mail: m.p.c.weijnen@tudelft.nl

© The Author(s) 2021
M. P. C. Weijnen et al. (eds.), *Shaping an Inclusive Energy Transition*,
https://doi.org/10.1007/978-3-030-74586-8_2

Introduction

On 11 December 2019, EU President Ursula von der Leyen presented the European Green Deal to the European Parliament. The European Green Deal embodies the ambition for Europe to become the world's first climate neutral continent by 2050. Quoting Von der Leyen in her address to the European Parliament: *"The European Green Deal is our new growth strategy—for a growth that gives back more than it takes away. It shows how to transform our way of living and working, of producing and consuming so that we live healthier and make our businesses innovative. We can all be involved in the transition and we can all benefit from the opportunities."* (EC, 2019a). Von der Leyen refers to the adoption of the European Green Deal by the EU College of Commissioners as: *"Today is the start of a journey. But this is Europe's 'man on the moon' moment."* (EC, 2019b). At the conference on the first European Climate Law, 28 January 2020, in Brussels, EU Executive Vice-President Frans Timmermans described the challenge of transforming a society entirely based on carbon to a climate neutral society that can function without carbon as a change of *'tectonic nature'*.

The big words used in the presentation of the European Green Deal mark it as a very ambitious policy document indeed. It is certainly remarkable in its breadth, as it covers all sectors of the economy, notably transport, energy, agriculture, the built environment and industries such as steel, cement, ICT, textiles and chemicals. Another remarkable feature of the European Green Deal is found in its recognition of the social and ethical dimension of the radical transformation ahead: *"The European Green Deal sets a path for a transition that is just and socially fair. It is designed in such a way as to leave no individual or region behind ...."*

In this chapter, we take an infrastructure perspective in exploring the challenges contained in the European Green Deal. Understanding the role of infrastructure is key to identifying the deep-rooted hurdles in the societal transformation ahead and to accepting how it will affect all of us. In the words of Frans Timmermans: *".... this is not just an economic change, this is not just a change in how we produce and live, this will affect every single institution upon which society is based and that helps society function as it does."* In his opinion: *".... the technology, the science, the money is not the problem, so why is it difficult? I think the essential issue is one of governance."* (EC, 2020a).

With a focus on the Netherlands, Frans Timmermans' home country, we will explore why and how the governance of infrastructure, and energy infrastructure in particular, is crucial to accomplishing the policy goals of the European Green Deal. Revisiting and potentially reshaping the current governance structures then is the next inevitable step that will determine whether and how the promises of the European Green Deal will be delivered by 2050.

Structure and Purpose of This Chapter

In this chapter, we will first focus on the nature of infrastructure, which extends far beyond its technical dimension. We therefore introduce a definition of infrastructure which includes the services provided through infrastructure, thus highlighting the social dimension of infrastructure. Indeed, it is only through the services it provides that we appreciate infrastructure. For example, society is not interested in a few extra kilometres of railway track as such. What matters to us is that we now can travel faster from A to B. Our appreciation will also depend on the price of the ticket, the comfort of the carriages and the timetable. All these aspects are a matter of governance. Only if we include the provision of infrastructure services, we can see how current governance affects the performance of infrastructure for society, in the sense that not all groups in society may be adequately served. At the same time, it is also possible to observe that some people may be disproportionately disadvantaged because their living environment is negatively affected by the external effects of the provision of infrastructural services to others.

Then we will sketch a picture of trends and patterns in the infrastructure landscape, which already have had a considerable impact on society. We argue that these should be accounted for in the reshaping of infrastructure governance in the EU and its Member States, if the European Green Deal is to deliver on its promise that no one will be left behind in the great transformation ahead. A dominant trend is the ever more intensive interdependency between infrastructural systems, which comes to light in the energy transition in particular. Other important trends are the digitalisation of infrastructure, the technological and administrative decentralisation of infrastructure development, and the metamorphosis of the formerly passive end user into an active player in the infrastructure system, in a new role as a provider of services.

We explore the interconnections between these developments and illustrate how they have an impact on our society. Our exploration signals a number of risks. Unless all citizens can participate and share in the benefits springing from changes in the infrastructural landscape, situations may arise that are perceived as unfair and where citizens affected may feel excluded. Impairment of social justice and inclusivity can erode support for socially desirable infrastructural transitions that require major investments. Moreover, there are ethical values at stake. Conversely, we also see that broadly shared feelings of social injustice can be a decisive factor initiating change. For instance, wide public consensus in the Netherlands about the plight of the citizens of Groningen, the Dutch province plagued by gas extraction induced earthquakes, made the Dutch Minister of Economic Affairs and Climate (EZK) decide to accelerate the phase-out of gas extraction in Groningen. It also persuaded many municipal governments in the Netherlands to phase-out natural gas altogether by 2030.

Obviously, social values are facts to be reckoned with in shaping the future of infrastructure. Nevertheless, in the practice of infrastructure policy, infrastructure is associated first and foremost with technical solutions, with economic value, with the support of economic activity, and with stimulating economic growth and productivity. In this chapter, we posit that this perspective does not acknowledge the rich social and cultural value which infrastructure creates for society. In this chapter, we shall attempt to interpret and delineate that social value, after having demarcated first what we understand by 'infrastructure'.

Note that this chapter is largely based on the state of infrastructure and infrastructure governance in the Netherlands. Between countries, even within the European Union, we see large differences in the details of infrastructure governance. The Netherlands, however, provides relevant illustrative material on the consequences of infrastructure policy. Firstly, the Netherlands ranks 4th in the 2019 Global Competitiveness Ranking Index 4.0 of the World Economic Forum, among others thanks to its excellent infrastructure quality (ranked 2nd in the world, after Singapore) (WEF, 2019). Secondly, it belongs to the group of most 'equal' societies in the world in terms of disposable income distribution among the population.[1] In the WEF report on the EU's progress on the way to achieving the competitiveness goals set in its 'Europe 2020 Strategy to achieve smart, sustainable and inclusive growth', the Netherlands ranked 4th for social inclusion (after Sweden, Denmark and Finland) (WEF, 2012). Yet, even in the Netherlands, social inclusiveness is not a given. As we will illustrate in this chapter, infrastructure policy decisions are of huge significance in that respect. We will advocate that infrastructure policy decision making should explicitly be assessed on its positive or negative consequences for the inclusiveness of society.

Defining Infrastructure

Infrastructure is an indispensable component of the human habitat in sedentary societies. Historical civilisations cleverly took advantage of the natural infrastructure of mountain passes, waterways and other vantage points when choosing places for settlement. The geography dictated where the demands of safety were compatible with the possibilities for subsistence and the needs for connection with the rest of the world. Moreover, the natural geography could be improved by waterworks, such as irrigation systems, fortifications and so forth. Today, the term infrastructure primarily brings up associations with man-made systems. In its original meaning, the term infrastructure was used to indicate the system of defensive works and military installations intended to protect society against enemy powers. As a river delta, the very existence of the Netherlands as a country hinges on the protective system of sea defences, river dikes and water level control. The meaning of the term infrastructure has gradually shifted to public works and amenities of general public interest, with an emphasis on their economic importance: networks for transport of people and goods, networks for electricity, fuels (for heating and transport), drinking water, sewerage, telegraphy, fixed-line and mobile telecommunication and data transport.

---

[1] See a.o. Eurostat, Gini coefficient of equivalised disposable income – EU-SILC survey, 15 December 2019.

In short, infrastructure involves the systems that accommodate the basic metabolism and signal processing of society in industrial and post-industrial economies.

While most infrastructure networks have changed hardly at all in the previous century, apart from significant expansion in capacity and density, radical technological developments have occurred in the telecommunications sector. Millennials and younger generations have no idea of things like a telex, a telegram or a fax message and no longer associate telephony with copper wires. Since the 1990s the telecommunications sector has seen a proliferation of new, digital, fixed-line and mobile networks, via which we communicate by means of speech, image, messages, data files, and so on. That development is not over yet. Alongside the development of the post-industrial economy and society we see the use of the term infrastructure shifting more and more to 'intangible' assets and services; terms like financial infrastructure, cultural infrastructure, and healthcare and knowledge infrastructure are common vocabulary by now. That indicates that our ideas about what services we consider essential are changing with the development of society and the degree of specialisation of the economy (Frischmann, 2012). That is not to say, however, that the infrastructure services that were once essential to the industrial society are not so anymore.

According to Wikipedia, *"infrastructure is the set of fundamental facilities and systems that support the sustainable functionality of households and firms, including the services and facilities necessary for its economy to function. Infrastructure is composed of public and private physical structures such as roads, railways, bridges, tunnels, water supply, sewers, electrical grids, and telecommunication (including internet connectivity and broadband access). In general, infrastructure has been defined as "the physical components of interrelated systems providing commodities and services essential to enable, sustain, or enhance societal living conditions and maintain the surrounding environment"*.[2] While it may be easy to reach agreement on a definition as provided in Wikipedia, the devil is in the detail when it comes to data collection.

Considering the crucial importance of infrastructure for the economy and society, it is surprising that basic data about infrastructure are largely lacking. Data about infrastructure e.g., about investments and capital stocks, are not consistently collected and if they are, it turns out that the data of different countries cannot be compared because of different definitions of infrastructure being used. Indeed, a common definition of 'infrastructure' is lacking in the international standards for compiling official macroeconomic statistics. Neither the 2008 System of National Accounts (UN, 2009), nor the European System of Accounts 2010 (EC, 2013) provide a clear definition of infrastructure and a clear delineation of what investment categories and assets to (not) be included. The struggle with a clear definition is also seen in the diverging definitions of infrastructure used in the United Kingdom, Canada and the USA in developing satellite or extended accounts on infrastructure. The definitions differ in the categories of infrastructure included. The UK Office of National Statistics sticks to a narrow set of six categories of physical capital assets, jointly referred to as

---

[2] Wikipedia "Infrastructure". Consulted 03–20-2021.

"economic infrastructure", where the flow of services or benefits accrues to multiple industries beyond the industry possessing the assets, like transport, energy, water, waste, communications and flood defences (ONS, 2018). The delineation used by Statistics Canada is much wider, including not only the aforementioned economic infrastructure assets, but also schools, colleges, universities and other educational buildings; libraries; hospitals; public security facilities; recreational facilities; memorial sites and so forth, which are often denoted as "social infrastructure" (Statistics Canada, 2018). What the UK ONS and Statistics Canada definitions of infrastructure have in common is that they focus on tangible assets.

The definition of infrastructure that we use in this chapter is in line with the UK Office of National Statistics' delineation of the so-called economic infrastructure: the collection of systems that protect us from flooding and drought, provide us with water and energy, communication, data and physical transport services, and ensure the hygienic removal of waste and wastewater. Thanks to that infrastructure, two thirds of the Netherlands' population can live and work in areas prone to flooding. Thanks to that infrastructure, we have energy and clean water at our disposal at all times and we can count on our waste and wastewater being discharged and processed hygienically. And thanks to that infrastructure, we are connected with each other and with the world around us.

In the way we define infrastructure in this chapter, however, we depart from the traditional 'hard infrastructure' definition. Rather than only referring to a class of physical and technical assets, we also include the intangible assets and systems required to produce the service providing functionality of infrastructure. For instance, a railway track as such cannot provide a safe and reliable public transport service, without access facilities, without appropriate carriers, and without traffic management and control systems. After all, the value of infrastructure manifests itself only through the services it provides for society and the economy. Hence, we define infrastructure as an essential services-providing socio-technical system. As all of the essential services provided by infrastructure hinge on a supply chain, or a set of interdependent supply chains, our infrastructure system definition thus covers the collection of supply chains which provide us with flood protection and water management, energy (electricity, heat, transport and heating fuels), transport of people and goods (overland, by air, water and rail), information and telecommunication (including digital communications and data transport), safe drinking water, sanitation and solid waste management.

The latter definition of infrastructure has been made operational by Statistics Netherlands in a validated method to estimate the contribution of infrastructure to the Gross Value Added (GVA) of the national economies of fifteen OECD countries and EU Member States over the years 1995–2016 (CBS, 2019). Despite huge differences between the fifteen countries in climate and geography, population density, spatial economy, economic structure, distribution of the population, and so forth, this study revealed that the share of infrastructure in the GVA of the national economy is quite consistently in the 10–15% percent range for all countries included in the analysis. For the Netherlands, throughout the years 1995–2016, infrastructure contributed 13.1%

(on average) to the GVA of the national economy. Critics may argue though that the infrastructure contribution as quantified in the Statistics Netherlands study still grossly underestimates the importance of infrastructure for the national economy, as ultimately all economic activity depends on the essential services provided by infrastructure.

Out of Sight, Out of Mind?

How Do We See Infrastructure?

In the Netherlands, as in most Western societies, we are used to take the availability of infrastructure for granted in planning and performing nearly all our everyday routines; at home, on the road and at our work place. We do not think about infrastructure, in the same way as we do not need to think about the functioning of the nervous system or the blood circulation in our bodies. In our sophisticated economy, we are hardly aware of the fundamental role infrastructure plays in the functioning of society, also because of the high reliability of infrastructure service provision.

Lack of awareness is caused further by the literal invisibility of much of the physical infrastructure, in the form of airwave frequencies and underground cables, pipes and ducts. In Western economies, most of those cables and pipes have been there for many decades and some of them much longer. Many of these assets have surpassed their technical lifetime and need to be replaced between today and 2030. That implies a massive investment challenge for the infrastructure network providers and for governments owning and managing infrastructure assets, such as the municipal governments in the Netherlands that run the sewerage systems. Whilst infrastructure renovation and replacement provide opportunities for innovation, their execution also causes great nuisance to end-users and local residents. In other words, when infrastructure 'comes to light', it is usually not a positive experience for citizens. And as long as the physical infrastructure assets remain buried underground, the motto seems to be: Out of sight, out of mind.

The linear infrastructure of roads, railways, waterways and high-voltage lines is always prominently present in our living environment, though. These links and the networks they are part of literally structure the spatial environment. The locations of roads, railways and waterways determine where we can live and work. The nodes in the networks of linear infrastructure are also nodes of business activities and social interaction. This is not only true for nodes in a narrow sense, such as railway stations, airports and ports. It also applies to nodes in a broader sense: cities developed historically around the nodes in infrastructural networks. Cities are nodes at the aggregated level of the infrastructure system-of-systems. Cities can exist thanks to infrastructural facilities and services. At the same time, successful cities also call for more and more infrastructural facilities to accommodate population growth and economic development. In many cases there is a self-reinforcing process of preferential connection with well-connected nodes (Barabási & Albert, 1999; Batty, 2008; Bettencourt et al, 2007). This process, also known as associative growth in networks, seems to account for the biased urbanisation patterns that we see in many countries, like the United Kingdom or France, where the urban agglomerations are dominated by one or a very few metropolis(es).

The distributed urbanisation pattern of the Netherlands is one of the few exceptions to the rules of associative growth. Historians diagnose the origin of this deviating urbanisation pattern in the fine meshed network of waterways in the Netherlands existing since the Middle Ages, and in later centuries supplemented with railways and motorways. This allowed the regional specialisation of the Dutch economy to develop at a relatively early stage, so that many comparatively small, connected cities developed, instead of one dominant metropolis (Van der Woud, 1987). In an international comparison, however, it would make sense, given the scale of the Netherlands and the relatively short travelling times, to regard the whole country as one coherent conurbation.

How Do We Experience Infrastructure?

Even when infrastructure is prominently visible, we often do not recognise it as infrastructure. Most infrastructure is deeply embedded in the spatial structure, both in the green and blue landscape and in the built environment. Thus, we experience a large part of the historical infrastructure as a self-evident part of the landscape and the urban environment. In the Netherlands, we can think of the rings of canals in many inner cities, the man-made dwelling mounds (terps) in areas prone to flooding, the star shaped fortifications with moats surrounding dozens of historical cities, and so forth. That historical infrastructure does not only create tourist attractions and cultural value. Many historical infrastructure assets still fulfil their original function, as do the canals which were constructed in the early nineteenth century by the Dutch king William I (nicknamed canal king). The function of the canals in our cities is being rediscovered as municipal governments are compelled to develop climate adaptation strategies. In several places filled-in canals are reconstructed, and in the development of new residential areas, canals and ditches often provide water storage and drainage functions. Most Dutchmen tend to forget that the quintessentially Dutch polder landscapes owe their existence, and their survival, to infrastructure, in the form of ring dykes and ring canals and pumping stations. The systems of river dykes and flood plains are infrastructure every bit as much as the dyke roads themselves. Indeed, only few Dutchmen will know that many of the nature conservation areas in their country originate in the historical energy transition from firewood to peat as the dominant energy carrier. Without historic peat extraction activities, many of the cherished lakes and landscapes in the Dutch provinces of Drenthe, Overijssel and North and South Holland would not exist (Van der Woud, 1987, 2020). On its website, the Dutch National Forest Service (*Staatsbosbeheer*) writes about a magnificent and seemingly pristine nature conservation area like the *Weerribben* in the north of Overijssel: *'Each meter of land here is man-made.'* Besides peat extraction, hydraulic engineering works played a major role in creating the land and the landscapes of the Netherlands which may be, more than any other country, a country defined by its infrastructure.

Historical infrastructure is often cherished as (pre-) industrial heritage, after its original function has been lost. This is true, for instance, for many old water towers, which are coveted today as residential and business premises, or for the old windmills and steam-powered pumping stations by means of which the Dutch polders were reclaimed in previous centuries. Throughout the world, there are countless examples of historical railway stations (e.g., Mumbai, Kuala Lumpur, and Dunedin) that are cherished as cultural heritage. Many of those stations are still functional, for that matter. Modern railway stations are often architectural highlights as well. Apart from their intended functionality they add aesthetic value to their environment, thereby contributing to a sense of place. The same goes for civil-engineering structures like bridges. Historical cross-river connections such as the Golden Gate Bridge and Sydney Harbour Bridge have become icons for San Francisco and Sydney, respectively, and in a similar fashion, the people of Rotterdam have embraced the Erasmus Bridge as 'their' swan.

Although modern infrastructure systems for drinking-water supply and sanitation are hardly visible, they are significant additions to the way we experience comfort, just like electricity and natural gas. Besides, hygienic drinking-water supply and wastewater discharge contribute significantly to the quality of the natural environment and public health. Thanks to these amenities, epidemics of cholera, typhus, dysentery and other water-related diseases are largely things of the past in our regions. In 2007 the readers of the British Medical Journal even elected the sewerage system as the most important 'medical' breakthrough since 1840 (BMJ, 2007).

How Do We Value Infrastructure?

Public health is of evident importance to the well-being of the population and thus brings significant economic value. The latter is not expressed, however, in the contribution of the drinking water infrastructure and sewerage to the gross value added of the Dutch economy. The contribution of drinking water, sewerage and waste management services combined to the Gross Value Added of the Dutch economy amounts to a mere 0.65% on average over the years 2006–2016 (CBS, 2019). Judging by the value added of the drinking-water infrastructure, it hardly 'pays' for the national economy to invest in that. It is evident that this reasoning does not take account of the actual function and value of a safe drinking-water supply for public health. For the energy supply sector,[3] the contribution to the GVA of the Dutch economy amounts to less than 2% (1.7% on average over the years 2006–2016). Yet it is evident that

---

[3]Covering the supply chains for electricity, natural gas and transport fuels (but excluding oil and natural gas production).

the crucial importance of energy services cannot be overestimated. Without energy services we would be miserable and the entire economy would collapse.

The current investment logic in the world of infrastructure is underpinned by a utilitarian perspective: infrastructure requires major investments, which are justified because infrastructure enables us to create more economic value. In this logic, benefits in the distant future weigh less heavily than benefits that may be realised in the short term. In the social cost–benefit analysis (SCBA) of infrastructural projects, aspects of non-economic value are represented only in monetary estimates. In today's practice this applies in particular to aspects of safety and public health—think of noise hindrance and air quality—, nature and landscape values. The current practice of SCBA falls short in acknowledging the rich variety of social and cultural values that infrastructure may create, like aesthetic quality and iconic value of prominently visible works of infrastructure. Such value aspects are not or hardly weighed in infrastructure investment decisions. In addition, the emphasis on (in) directly quantifiable economic benefits and their quantification in the SCBA system suggests that the economic benefits of infrastructural investments can, and should, be predicted, as a justification for building the infrastructure.

In this context it is a sobering fact that the relationship between investments in infrastructure and economic output still has not been clearly established scientifically. This is borne out among other publications by a meta-analysis of 80 macroeconomic models by the World Bank (Straub, 2008). Despite the well-recognised role of infrastructure as backbone of the economy, its exact economic value is difficult to determine (Aschauer, 1989; Carlsson et al. 2013; Munnell, 1992). From a macroeconomic perspective, it is clear that infrastructure contributes to economic development. In the Global Competitiveness Index, which is published every year by the World Economic Forum (WEF), the role of infrastructures changes along with the development level of the economy. According to the methodology applied by the WEF, in factor-driven economies the quality of the traditional infrastructural basic facilities accounts for 25% of the competitiveness score; this specifically concerns roads and railways, shipping and airline infrastructure, electricity infrastructure and networks for fixed-line and mobile telephony (WEF, 2017). It is easily understood that economic development is hardly possible when such infrastructural basic facilities are lacking. In more advanced efficiency- and innovation-driven economies, the competitiveness index calculations of the WEF indicate that the relative importance of such basic facilities decreases. In contrast, the accessibility and quality of fixed-line and mobile internet and communication facilities assumes a greater role. Then, the importance of infrastructure also declines in favour of other competitiveness factors, such as the efficiency of goods, labour and capital markets, the quality of the knowledge infrastructure, and the legal system.

From a microeconomic perspective, investments in infrastructure have direct and indirect effects. In general, the direct effects, for example as related to travel time reduction, can be quantified well (Romijn & Renes, 2013). By contrast, the indirect economic effects of infrastructure are far more difficult to prove. These may be business location factors, such as proximity and agglomeration effects, the increase in value of real estate, the growth or decline in local activities due to relocation and changes in commuting behaviour, and the effect of image—think of hubs and hotspots. Indirect effects of the implementation of infrastructure may also concern social effects, which are even more difficult to quantify in monetary terms, if at all. Many examples can be given of infrastructure investments which, contrary to the prognosis, did not yield a return or only yielded a return many years later than planned, and of investments which turned out to be far more profitable than anticipated or profitable in another way than anticipated. Recently the Netherlands Bureau for Economic Policy Analysis (CPB) reported that, in less than two years after the construction of the A2 highway tunnel in Maastricht, the increase in value of the building stock within a one kilometre-radius from the tunnel amounts to some 220 million euros. This is nearly twenty times higher than what was estimated ex ante. And this does not even include other quality-of-life benefits and the travel time reduction effect (CPB, 2018a, b). Especially indirect effects of infrastructure investments are surrounded by a great deal of uncertainty. It must be noted though that even the envisaged direct effects are often only realised very late. In the Netherlands, notorious examples are the Betuwe line (a dedicated freight railway line) and Groningen Seaports. In retrospect it becomes visible, however, that major infrastructure investments have usually had a tremendously positive impact on the relevant regions, even if they worked out differently than initially envisaged. Even though we still cannot properly foresee whether and how infrastructure investments will yield a return, we do know that the absence of infrastructure is a guarantee for socioeconomic development failing to occur. Indeed, all of the Sustainable Development Goals on the United Nations agenda hinge on investments in infrastructure (Thacker et al., 2019).

Moreover, in making investment decisions today, we should be aware of the fact that infrastructure investments throw a long shadow into the future. In accounting for these long-term effects, it helps to remind ourselves how decisive the infrastructure investments of our ancestors still are for the country in which we live and work in this day and age. The entire infrastructure in operation today represents a huge capital, which was largely invested in the past. Yet, infrastructure investments in the past were made for another society, with another economic structure and other societal priorities than we have today. Most of the Dutch canals, railways and ports were constructed deliberately to foster the development of large-scale industry, bulk transhipment and transport, in order thus to create employment and strengthen the international trading position of the Netherlands. The discovery of natural gas in the province of Groningen not only led to a rapid introduction of natural gas as a clean replacement for coal and petroleum in Dutch households and businesses in the 1960s; it also motivated the deliberate attraction of energy-intensive industry to the Netherlands. The Groningen natural gas reserves thus contributed significantly to the improvement of air quality as well as economic prosperity in our country. In hindsight, it may be said that this was achieved at great social and economic cost for the people in Groningen, who are now suffering from gas extraction induced earthquakes. Still, it is too easy to say that, with today's knowledge, those historical choices could not have been accepted as sustainable. Even if we assess these historical investment decisions as unsustainable, that does not imply that all of this historical infrastructure is now obsolete. Most of this infrastructure can be adapted for renewable energy carriers and sustainable processes in the future while adequately serving us today. We should also acknowledge that our views of sustainability have changed drastically and, like the generations before us, we cannot read the future either.

Given the evident uncertainties, it is good to note that the physical sustainability of infrastructure investments, which are deeply embedded in the spatial and economic structure, does apparently not impede a society in flux. Despite its physical inertness, established infrastructure has so far proven to be able to support a constantly changing society. A lot of old infrastructure still represents great economic and social value. However, over time we see a change in the values that society wants to create with infrastructure and the values it wants to enshrine in topical infrastructure development projects.

Traditional Values in Infrastructure Systems and Services

Traditionally, infrastructure makes us think of public amenities: provided, or regulated, by the government; the costs of which are socialised, because they benefit everybody or because nobody can be excluded from benefiting. Military defence works and flood defences are evident examples of such public amenities. Gas and electricity supply, drinking-water supply, public transport services and fixed-line telephony are examples of amenities which were brought about at a local scale by private initiative, but were soon taken over by the government with a view to economies of scale, characteristics of natural monopoly (with the appurtenant risks of abuse of market power) and positive network externalities in conjunction with public interests. Today, the collective nature of those amenities is no longer considered self-evident, though.

It is a moot question whether we would ever have realised the universal access to drinking water, sewerage, electricity and natural gas, which we take for granted in the Netherlands, if the neoliberal paradigm, which has been in force since the early 1990's, had been leading in this pursuit. In many developing countries, we see that such infrastructure facilities and services are provided only in urban agglomerations, whereas the unprofitable connections to and within rural areas are not forthcoming. In the Netherlands, the principle of universal access to many infrastructure services is legally enshrined in the form of connection rights and obligations for drinking water, sewerage, electricity and, until 1 July 2018, natural gas. Furthermore, there are legally established quality requirements, intended among other things to guarantee safety and public health. All of these measures have been of significant importance for the high quality of infrastructure provision in the Netherlands. Meanwhile one may wonder whether access to fast internet has not become just as essential as electricity and water, and whether this implies that effective statutory obligations should ensue in this respect.

Access to essential services has not been realised equally well in all western economies, as we can see in countries like the United States. There we find persistent differences in the accessibility and affordability of essential infrastructure bound services between urban and rural areas, and between neighbourhoods with a high and with a low socioeconomic status—a separation that often runs parallel to ethnic lines. By no means all Americans get safe drinking water, in the countryside more than a quarter of the population does not have access to fast internet, and 85% of Americans cannot reach work, a hospital or shops without a car (Tomer, 2018a). Households in the lowest income quintile are forced to spend over 60% of their net income on drinking water, sanitation, electricity, gas, telecommunication and transport services. With housing costs added, the lowest income group has literally not a cent left for other necessities of life (Tomer, 2018b).

In the Netherlands, a conscious political choice for socialisation of infrastructure costs was made in the past to ensure the affordability of infrastructure services for all citizens. Nonetheless, in the Netherlands, too, there is inequality in access to and affordability of infrastructure services between citizens, between regions and between urban and rural areas. This is not a novel phenomenon. Differences between regions can be explained to a considerable extent by geographical conditions, such as the natural infrastructure of navigable waterways. In the topology of road, railway, electricity and gas networks, it is still visible that the so-called Randstad area[4] and other economically dominant urban regions were given priority in infrastructure development over the periphery. In the past, defence policy considerations played an important role in national infrastructure development decisions, particularly so with World Wars I and II still fresh in the collective memory. Later, economic profitability considerations (following the SCBA methodology) became dominant, putting a halt to daring ventures such as high-speed railway line (HSL) connections to the furthest corners of the country, incl. Groningen (Mouter et al., 2013).

In short, the dominant traditional values embodied in the design of infrastructure and infrastructure service provision can be summarized as (universal) Access, Affordability, Availability and Acceptability. Below we give a quick review of how these values are upheld per infrastructure sector in the Netherlands.

Information and Telecommunication Services

Thanks to technological innovation, the fixed telephone line met with competition from different forms of mobile telephony, which are no longer characterised by prohibitive costs for new providers wishing to enter this market. The natural monopoly of the fixed telephone line has come to an end as a result of competition with and between new fixed and mobile data networks. The same goes for digital information, which runs largely via the same networks. There are still quality differences between those networks, which manifest themselves mostly in terms of bandwidth and speed of data traffic. Providers of fast digital infrastructure must first recoup their investment before rolling out the new generation of telecom networks.

---

[4]The western part of the Netherlands spanning the cities of Amsterdam, Rotterdam, The Hague and Utrecht.

Moreover, they then seek a great density of connections. Consequently, the fastest service via a fixed fibre-optic network is still only accessible to a limited extent, depending on the region and municipality.[5]

In the 2018 Digital Economy and Society Index (DESI) of the European Commission, the Netherlands ranked first in Europe on the indicator of digital connectivity (EC, 2018),[6] with the Dutch top score for connectivity based especially on access to fast internet (72% of all households). In 2020, the Netherlands' performance had already fallen to the sixth position in the DESI 2020 ranking (EC, 2020b). Whereas the Netherlands still holds the top position in fixed broadband penetration (with 98% of households having a fixed broadband subscription), it is lagging behind other European member states as regards the penetration of ultra-fast internet in households. With respect to ultra-fast internet, Sweden is the front-runner in the EU, with almost 65% of households, thanks to spectacular growth in fibre to the home connections. The Netherlands seems to be slowed down by an inhibitory head start effect.

For entrepreneurs and citizens who cannot or not yet secure an optical fibre connection, this means that they are more limited in their options for business and personal development in the digital world than others who do have such connections. At present incremental innovation (Very-high-bitrate Digital Subscriber Line—VDSL—and bonded VDSL) can still improve the capacity of the existing copper network, as is true also for coax cable networks (with Docsis 3.1). Still, practice so far shows that new capacity is soon filled by higher user requirements: data use is increasing exponentially. This makes it questionable whether incremental innovation of existing networks provides enough new capacity to include all households even in the short term in data-intensive pricing systems (real-time or time-of-use) via smart grids, as is assumed in the energy transition.

Gas and Heat Provision

In the Netherlands, network providers have a statutory obligation to connect each household to electricity. Until July 2018, the statutory obligation to connect also applied to the natural gas network. For gas, an exception was possible only in case of an alternative connection to a heat network. Since April 2018, however, new dwellings will not be connected to the gas network anymore. The Municipal Executive may make exceptions in the event of compelling reasons of general interest, however. The manner in which heating services are provided, without natural gas, is left to the discretion of the municipal authorities, and private homeowners. At the national climate summit in October 2016, 77 Dutch municipalities signed the

---

[5]For the Netherlands, indeed, Bureau Stratix reported that the roll-out of optical fibre has stagnated for several years now.

[6]The indicator of connectivity measures the deployment and quality of broadband infrastructure, and considers both the supply and demand side of that infrastructure. This concerns fast (at least 30 megabits per second) and ultra-fast data connections (at least 100 megabits per second), both fixed and mobile.

manifesto 'Setting to work with living without natural gas' ('Aan de slag met wonen zonder aardgas'), in which they declared to abandon natural gas as a heating fuel ultimately by 2030.

This is far from easy, given the fact that at this moment more than 90% of households are still connected to the natural gas network. Given the time horizon of 2030 that many municipal governments set to abandon natural gas, they do not see hydrogen as a viable alternative. Neither green nor blue hydrogen will be available in sufficient volumes and at an affordable price before 2030. Moreover, most likely, a primary candidate for the use of (initially scarce) green hydrogen will be industry, which has a very limited set of alternatives to generate high temperature process heat. Municipalities are focusing on alternative options for space heating that can be implemented sooner. However, there are great local differences in the options for municipalities to provide a sustainable alternative. In agricultural areas and in the vicinity of waste water purification plants there is a potential supply of biogas. Residual heat is abundantly available near industrial parks with a lot of process industry. The possibilities for heat extraction from surface water or for geothermal heat extraction are not the same everywhere in the Netherlands, and there are considerable differences between districts in the energy quality of the built environment. Energy neutrality is in the process of being defined as a statutory obligation for new premises and more and more new housing estates are already constructed according to standards of near-energy neutrality. Energy neutrality, however, is an expensive requirement for older buildings and houses. In the event of renovation, the heat requirement can be reduced substantially, but this comes with a hefty price tag. Given the different options at different locations in the Netherlands, it remains to be seen how the costs and the security of supply of heating services will develop in a natural gas-less future. Whereas today the principle of 'No More Than Otherwise', using natural gas as a reference, is still applied in the heat supply tariffs, it is not inconceivable that considerable differences in costs will occur in the future per region, per municipality and per district.

Electricity Provision

For electricity end-use an alternative is not under discussion. The consumption of electricity is only rising due to the increasing need for household comfort in an ageing society, due to the rapid advance of digitalisation—think of datacentres- and due to electrification of energy functions that were formerly fulfilled by other energy carriers -think of electric heat pumps and electric vehicles. The huge change taking shape here is the production of electricity from renewable energy sources, sun and wind in particular. These are partly large-scale developments, like the wind farms on the North Sea, while small-scale production is rapidly gaining ground as well. Many farmers already use their land, farmyards and buildings for wind turbines and solar PV parks, and increasing numbers of private homeowners put solar panels on their roofs, encouraged by attractive subsidy schemes and feed-in tariffs.

Consequently, we see more and more signs of inequality emerging in the supply of electricity from renewable sources, as the possibility to exploit their assets is available to landowners and homeowners, but not to tenants. Considering that it is not the poorest citizens who possess the financial resources to have solar panels installed, and that such activities also require due knowledge and 'acting ability' (WRR, 2017), we see a possible future scenario unfolding here in which inequality between citizens in the affordability of electricity, an essential service, increases. It is up to housing corporations to make the energy supply for their tenants sustainable. However, there are great differences between housing corporations, among others in the availability of financial resources. This implies, in practice, that different housing corporations do not serve tenants in a similar manner.

A scenario of increasing inequality in the electricity supply is becoming even more likely when we realise that the large-scale introduction of weather-dependent energy sources will inevitably lead to greater volatility on the supply side, whilst there are ever fewer traditional, coal- and gas-fired power plants which can be regulated up and down easily. This leads to rising pressure to bring about flexibility on the demand side. In industry a lot of flexible demand is available which is already being used cleverly to respond to price fluctuations in the wholesale market. Nonetheless, in the future an appeal will be made to households as well for flexibility in electricity demand. At this moment, the electricity demand in an average household is hardly elastic, but that may change in the future, as more households will have electric heat pumps and electric cars, and storage options for electricity and heat. There are incentive schemes available for such facilities, in the form of subsidies and easing of the tax and premium burden. Similar to investments in solar panels, it is a given that those benefits end up mostly with the relatively prosperous and highly educated section of the population that own their residences.

Other Infrastructure Services

It is less evident whether a trend towards unequal treatment of regions or citizens is coming up in other infrastructure systems as well. As in many other countries, the spatial demography is changing, as more and more people leave the countryside for better job and education opportunities in urban agglomerations. In rural areas where the population is shrinking, this may have an impact on the quality of infrastructure services. Provinces and municipalities are already contending with the organisation of adequate public transport facilities for a population of on average older and less mobile citizens. As a result, there are many locations in rural areas where only a very low frequency bus connection is offered, in some cases even manned by volunteers. The financial service provision by bank branches and ATMs and the social infrastructure such as schools, fire brigades, ambulance services, hospitals and other healthcare provisions are under pressure in those areas as well. As and when those facilities become scarcer, the mobility demand becomes greater and more urgent.

In general, less attention is devoted to the consequences of these demographic developments for the more basic service of drinking-water supply: As the user population in a region is dwindling, there is a threat of having an over-dimensioned drinking-water infrastructure. This may have a negative impact on drinking-water quality, unless investments are made in additional monitoring and possibly even in physical adjustments to the distribution network.

Another challenge, as an example, is that the combination of an ageing population and the current policy to treat sick patients outside the hospital as much as possible, and to let elderly people live in their own homes longer, will lead to an ever-heavier burden of household waste water with antibiotics and other medication. Today's sewage treatment plants are not equipped for the removal of such compounds and are facing new investments. In this case, there is actually little inequality between urban and rural areas, between more and less prosperous citizens, and between regions. In all regions, sewage treatment plants will need to be adapted so as to purify the wastewater from these new contaminations.

As we can see, the traditional core values of infrastructure (service) provision are still very much alive. The manner in which these values are upheld, however, has drastically changed with the reform of the traditional public monopolies that used to supply these infrastructure services, and still do so in the case of drinking water supply, sewerage, flood safety and wastewater processing. Following the introduction of market forces into other infrastructure sectors, we observe tendencies towards more inequality in the availability, quality and affordability of essential infrastructure services, like the supply of digital information, energy and transport services. This particularly affects users in rural areas, where profitable or even cost-efficient service provision is more challenging. It also affects the less affluent users, especially those who cannot benefit from financial incentive schemes for home improvements (thermal insulation), solar panels, heat pumps, electric cars and other measures designed to support the transition towards a sustainable energy system.

In the following sections of this chapter, we will focus in more detail on topical developments in the infrastructure sectors, especially in the energy infrastructure. We will identify some emerging coordination issues and examine the challenge of energy infrastructure governance in safeguarding the core values of infrastructure (service) provision in the years to come.

Decentralisation

Many people associate the energy transition with the rapid increase of installed capacity to harvest energy from renewable energy resources that they observe in their living environment. Wind turbines and large-scale wind parks have become prominent features in the landscape and solar roof panels are other visible signs of the energy transition. These new energy technologies are omnipresent unlike the large-scale fossil fuelled power plants of the past. The relatively small scale of many renewable energy technologies has drastically lowered the entrance barrier to the power generation market and thus enabled many new actors, including individual citizens, to enter this market. Indeed, government has actively encouraged citizens to engage in the use and self-generation of power from renewable energy resources by offering a range of stimuli e.g., information campaigns, subsidies, fiscal incentives and attractive feed-in tariffs. In the same vein, government has developed incentive schemes for home improvements, heat pumps and energy storage units. At the same time, new building standards are imposed, especially with respect to thermal insulation, and all homeowners who want to sell their property are required to apply for an energy label.

The smaller scale of renewable energy technologies is inviting to citizens who want to join in the combat of climate change as well as to citizens who are keen to reduce their energy bills. In an increasing number of neighbourhoods, energy cooperatives have been and are being established. The Netherlands counts a total of 582 energy cooperatives at the end of 2019, an increase of almost 100 new cooperatives since the previous poll in 2018 (HIER, 2020). These cooperatives were established for the goal of energy saving and collective power (and sometimes heat) generation, with the additional benefit of community strengthening. It turns out that cooperatives hardly engage in large-scale projects because these come with high uncertainty about the future tax incentive regime and thus with high uncertainty about the cost-effectiveness.

Whereas power generation used to be a capital and knowledge intensive activity reserved to highly specialised actors, the energy transition is bringing it down to the level of local communities and individual citizens. The ongoing decentralisation of power generation is enabled by the established power transmission and distribution infrastructure, which accommodates the natural variability of renewable energy resources. Seen from the solar roof owning citizens perspective, the electricity grid is the battery that absorbs temporal surplus production and supplies during times when generation falls short (Bakker de, et al., 2020; Hufen & Koppenjan, 2015; Verkade & Höffken, 2019). In many respects, this situation is comparable with the current practice of spatial heating in the Netherlands, where homeowners operate their own individual 'central' heating installation fuelled by natural gas, with natural gas being supplied through a fine-mesh grid with nationwide coverage. Here, the energy transition calls for a shift away from natural gas as a heating fuel. Many municipal governments are now in the process of implementing heat networks, especially in densely populated urban areas.

The energy transition does not only bring radical change in technology and physical structures, but also applies to new rules and coordination mechanisms for the parties that realise the provision of infrastructure and infrastructure services in a concerted action. Examples are the design of energy markets and network regulation, or the re-allocation of roles between the public and private sectors, and between the central government, the European Commission and local and regional authorities. We see that political decision-making has in recent years led to a pattern whereby the specifics of various policy domains, including infrastructure policy, are increasingly determined at the level of municipalities. The national government sets relatively broad frameworks, while the role of provincial authorities is limited. It is due to the network nature of most infrastructure systems that local infrastructure often forms part of regional, national and international systems, up to a continental or even global scale.

Now that the responsibility for infrastructure development is shifting more and more to local and regional authorities, this raises questions about the coordination between scale levels in the system of infrastructure. Development choices that are rational at a local level may be at odds with desired developments at national and international levels, and vice versa. The multi-scale character of infrastructure is not a new phenomenon, but conflicts of interests came to light less harshly when the top-down coordination by the government was still a matter of course. Now that the lead is simply thrown onto the desk of local authorities, the coordination issue becomes more pressing. Indeed, the coordinating and administrative capacity of municipalities is limited and differs per municipality. It also remains to be seen to what extent the present system of democratic representation in the municipal government is appropriate to shoulder the future administrative burden. Furthermore, the administrative decentralisation of infrastructure policy to the municipal level means that new risks arise of a potential increase in inequality in the provision of infrastructure, as a result of policy competition between municipalities and between regions.

Digitilisation of Infrastructure

In all future visions for electricity supply, there is a vital role to be played by active demand response, induced by time- and location-dependent price signals, in accommodating the variability of electricity production from renewable sources. In view of the short response times required to effectively utilize demand flexibility, this can be facilitated by automating that demand response. This is where the so-called smart grids come in, which enable digital communication between the electricity end-user (and end-use equipment) and the network operator. For other forms of energy supply, too, there is mention of intelligent networks that can support this active demand response, although the need for intelligence in heating networks or networks for 'other' gas e.g., biogas, hydrogen and synthesis gas, will be less pressing as indeed, there are realistic possibilities for storage.

Active demand response to time- and location-dependent price signals is an option, which has also been discussed for many years for road use within the context of combating traffic congestion. However, there it has so far resulted only in small-scale realisation for certain road sections. The kilometre charge for trucks announced in the Netherlands in the 2017 Government Coalition Agreement is not dependent on time or location, but is only differentiated according to emissions, in conformity with the model already being applied in other European countries e.g., Germany and Belgium. Up to now a national, more sophisticated system of kilometre charge for both passenger and freight traffic to mitigate congestion by means of time- and location-dependent stimuli has met with massive political resistance, even while public resistance is dwindling. Dutch Railways has applied a difference in train fares between peak and off-peak hours and according to the age of travellers. Still, most commuters do not regard this as a free choice, but as an inevitable peak-hour charge.

The introduction of active, momentary demand response is an essential change in the access to infrastructure, which will have a profound impact on the manner in which and the time at which citizens use the services provided by infrastructure. This may have a great impact on the organisation of their daily activities. A precondition for facilitating active demand response is a far-reaching digitalisation of infrastructure networks and end-use devices. Through the introduction of smart meters, remote monitoring and time-dependent pricing, the electricity supply is increasingly becoming an integrated system of IT, telecommunications and electricity supply. In the future, this amalgamated system will integrate with electric means of transport. Digitalisation is also transforming the infrastructure systems for transport (Uber), public transport, telecom, radio and television, enabling among others the provision of services on demand. Digitalisation, however, gives rise to a number of concerns.

Firstly, there is the reliability of the system. Although the network providers use self-administered IT and telecom systems independent of the public internet as much as possible for their own operational and management tasks, cyber vulnerabilities are unavoidable in the connections of electricity users with their energy suppliers. It is self-evident that these interfaces present vulnerabilities that may be exploited with malicious intent to disrupt society.

Secondly, there is the nature of the companies managing the data platforms. Making IT-controlled demand response operational requires detailed insight into the daily activities and personal preferences of citizens, for a large number of parties and platforms involved. A question that arises in this context is how the new digital platforms, which enable citizens to move in the market as producers and sellers of electricity, digital content, mobility services and such, should be regarded—and regulated—as part of the infrastructure system. The enabling data platforms are created or managed by companies that do not themselves contribute any infrastructural hardware assets. They 'merely' provide a platform to intermediate between supply and demand and thus enable private asset owners to exploit their assets commercially. This raises the issue as to whether the regulatory treatment of such data platforms is unfair where it concerns the enabling of retail services that compete with those provided by traditional firms, which are stuck with the maintenance of capital-intensive infrastructure. The positive network externalities of those digital platforms are so huge that the relevant companies have developed at unprecedented speed into virtually global monopolists, such as Uber, Airbnb and Google. At present, their business model is based on gathering and analysing data about our behaviour in data centres, with a specific location under a national jurisdiction. In the future, this will take place increasingly in a distributed cloud, on a large number of different systems distributed across the globe.

Although such companies are not giving rise to a natural monopoly, as is the case with capital-intensive physical infrastructure assets such as railways or electricity networks, but depend on positive network externalities, that does not detract from their market power. The risk of abuse of market power in a natural infrastructure monopoly can be curbed by ex-ante regulations. For globally operated data platforms, however, this kind of regulatory control is more complex. First and foremost, because these platforms operate in different jurisdictions. Besides, effective regulation requires access to and understanding of the specific algorithms which these platforms use to analyse the data and to manage the connected 'customers' in their transactions and behaviour. Even ex-post monitoring, by virtue of competition legislation, is very difficult and moreover hardly effective: then the damage has already been done.

Thirdly, there are ethical aspects at stake. The power of such platforms is that they facilitate a massive volume of transactions between connected customers, suppliers and service providers. Hence, they get to possess a great deal of information about the connected parties. Through clever analysis of the data, they can bring these different parties into contact with each other and make them enter into the most advantageous transactions. This means that supply and demand can be attuned to each other better and that the parties have far more options to realise their freedom of choice of products, suppliers or customers. Thereby such a platform also provides the option to discriminate very effectively, depending on the characteristics of suppliers and customers. It just depends on the kind of analysis and selection algorithms deployed and on who is given what information, and under what conditions. The degree of 'impartiality' of the platform is decisive for the way in which the information will or will not be used to discriminate between groups of users, on whatever grounds. Platforms can be smart and benign or smart and mean, depending on their business models.

Infrastructure Interdependencies

Apart from the organisation of vertical coordination between spatial and administrative scale levels within specific infrastructure networks, the increasing interconnectedness and interdependencies between infrastructure systems have also created new coordination issues. The dependencies between different infrastructure systems are big as they are and are becoming ever more critical, amongst others due to the progressive digitalisation in the operation and maintenance of physical infrastructure, in the operation of infrastructure markets and in the actual provision of infrastructure services. Digital infrastructure is already an indispensable part of electricity infrastructure as, indeed, it is in transport infrastructure. On a higher level, all infrastructure critically depends on energy infrastructure services: it is impossible to ensure water safety, safe drinking water, transport of people and goods, telecommunication and data services without a reliable supply of electricity and other energy carriers. In turn, data and telecommunication infrastructure relies on electricity to power masts and data centres. The processing of waste and of wastewater is closely connected with energy supply and demand, and energy supply is impossible without transport of goods and data. While these cross-sector interdependencies are not new, they are steadily evolving towards an intricate network of myriads of deeply embedded interconnections; to the extent that a cross-sector amalgamate system is emerging. Especially energy, transport and digital infrastructures are thus being fused together into a new complex system.

Far-reaching integration of energy, IT/telecommunications and mobility infrastructure means that the traditional 'walls' between those systems and their current sector-by-sector and network-specific regulations are not tenable anymore. In this context, it is telling that the Dutch Ministry of Economic Affairs needed an order in council to set up test beds for intelligent energy networks, in which the allocation of roles between market parties deviates from what is permitted within current legislation and regulations.[7] Also heated discussions arose about the allocation of roles between network providers and energy suppliers in the roll-out of charging points for electric vehicles and in the realisation and management of energy storage facilities for temporal surpluses of electricity produced by (both central and decentral) renewable sources. Today's legislation dictates that such new activities must be provided by the energy service providers in competition. Thereby, it denies the public energy network managers important options for a cost-efficient development of their networks and more efficient use of available network capacity. In fact, the coordination of interactions between networks and services for energy, telecommunication and transport still forms a gap in legislation and regulation.

The current structure of infrastructure governance and policymaking is organised by sector and, within sectors, by infrastructure. This siloed structure lacks mechanisms that do justice to the ever more intricate interwoveness of the infrastructural system-of-systems. It does not take account of direct effects of policy measures in one infrastructure domain on other infrastructure domains, and of indirect effects on society. In the Netherlands, the introduction of a new Environment and Planning Act [*Omgevingswet*] seeks to provide this horizontal coordination, albeit with a narrow focus on the physical environment. Although social aspects can be observed and incorporated into the development of regional and local Environment Visions and Environment Plans, the Environment and Planning Act does not make this mandatory. Besides, as indicated above, it is also a matter of concern whether, particularly at the level of local and regional authorities, there is sufficient well-organized capacity and knowledge available to fulfil the potential of the Environment and Planning Act. In this respect, major differences between competent authorities can in due course also be a source of inequality between cities and regions in the supply and quality of infrastructure services. Another matter of concern is the time horizon of decision-makers versus the lifespan of the infrastructure. Is there enough room to think beyond the needs of today?

---

[7]Decentralised Sustainable Electricity Generation (Experiments) Decree, published in the Bulletin of Acts and Decrees (Staatsblad van het Koninkrijk der Nederlanden) no. 99, 10 March 2015 (only in Dutch).

New Challenges for the Governance of Energy Infrastructure

Multi-scale and cross-sector interactions within and between infrastructure systems are a given in today's infrastructure world. The big change in historical perspective is that the coordination of these interactions has become both more critical and far more complicated. It is more critical since modern post-industrial economies, with their many long and specialised value chains, will simply collapse if the flows of data and energy are interrupted. It is more complicated as today's infrastructure is a kaleidoscopic (inter)national system with some subsystems residing in the public sector, some in the private sector and others in semi-public and semi-private realms. As the latter complication also occurs within infrastructure value chains, the uninterrupted supply of essential infrastructure services depends on the adequate coordination of transactions across many interfaces.

In the energy infrastructure sector, the constellation of public and private actor involvement widely differs by country, even within the European Union. While the European Commission imposes strict institutional requirements for the operation of physical infrastructure and infrastructure bound markets, the Member States still have a considerable degree of freedom in shaping their national energy sector. At the same time, the physical reality is that all power transmission grids in Europe are interconnected across national borders, including the transmission grids of non-EU Member States. The multi-scale complexity of electricity and gas infrastructure, physically and economically, implies the involvement of multiple national, regional and local governments, besides the European Commission. The governance of these systems steers their performance, and the consequences are felt by national economies and by individual citizens alike.

Coordination of Transactions in Energy Infrastructure

At the heart of the governance challenge is the question of how transactions across the energy value chains can best be coordinated. For this question, we refer to the seminal work of Oliver Williamson on transaction cost economics and efficient contracting, for which he was awarded a Nobel Prize in 2009. Williamson decomposes the value chain into bundles of transactions between different actors with specific roles in the system, which require certain investments. Relevant characteristics of the transactions are, first, their specificity, second, the extent of uncertainty in the transaction environment and, third, the costs of creating safeguards (to curb uncertainty) relative to the transaction value. As shown in Fig. 1, Williamson distinguishes seven configurations with respect to these transaction characteristics, where each configuration calls for a specific form of coordination.

Configuration A denotes a transaction that does not require very specific and/or highly specialized investments that would create strong interdependencies between actors. This type of transaction can best be coordinated in a competitive market,

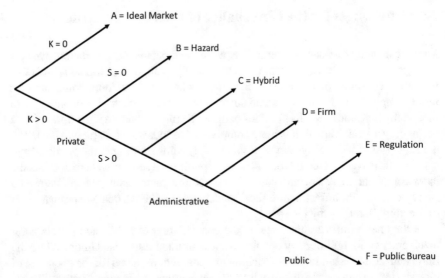

Fig. 1 Williamson's contracting scheme. *Source* Adapted from Williamson (1998). *Note* k represents a measure of transaction-specificity of the assets involved, while s denotes the safeguards needed to protect investments for transactions

if that market is characterised by sufficient numbers of actors on both the supply and the demand side. Configuration B represents a transaction that requires specific investments, creating a potential hazard for one or more actors, but where the level of uncertainty is fairly limited, so that contractual risks can be accounted for in the price setting in the market. Configuration C is concerned with a higher level of uncertainty (than configuration B), where it is assumed, however, that actors have sufficient insight into the nature and consequences of uncertainties to be able to create adequate contractual safeguards. In configuration C, these safeguards will typically involve long-term risk management, as a hybrid solution between the market and full integration. Configuration D involves transactions with high risks and high potential benefits, which are fully integrated and equally shared between the private parties involved in the transaction, for example, in a joint venture. Transactions in configuration E are characterised by high uncertainty and by large asymmetry in information and risk exposure between private parties. In this configuration government involvement is needed as an independent regulator and/or arbiter, setting and enforcing the market rules for such transactions. In configuration F, the uncertainties and associated risks are of such magnitude that no private investors are willing to take those risks in return for a societally acceptable fee. In this case, only government remains as a potential provider of the good or service. Configuration F thus leads to the establishment of a public entity that bears all the risk.

We can see the logic of Williamson at work in the way the neoliberal reform of the electricity sector has taken shape in the Netherlands, and in the European Union at large, since the 1990s. The production of electricity and the supply of electricity services are both seen as type B and C transactions, with a sufficient number and diversity of actors at both the supply side and the demand side of the competitive spot markets, or engaging into longer-term contracts. The electricity market operates under the scrutiny of regulatory oversight as regards price formation and abuse of market power. In contrast, the transmission and distribution segments of the value chain, given their natural monopoly characteristics, cannot be provided in a competitive market. For these segments a hybrid solution E was chosen with regulated private or public companies (national and regional monopolies) operating the networks. The network companies are heavily regulated with a focus on cost effectiveness, which enhances the affordability of electricity service provision.

With respect to the provision and creation of new network capacity, the network providers are expected to follow what the market needs with respect to transport capacity, rather than proactively enabling or incentivising particular market developments. Indeed, as enshrined in the EU electricity and gas directives, their primary task is to provide least cost solutions for the transport of energy from the producers to the consumers. The competitive commodity transactions dictate the need for transport capacity and its locational structure. The focus on efficiency in the regulation of the transmission and distribution networks, the complex procedure in awarding new investments and the consequent risk averse character of the distribution system operators create a relatively slow process of adjustment of the infrastructure. This explains why investments in transmission and distribution network capacity are lagging behind the much faster development of installed power generation capacity from renewable resources. For example, in many rural parts of the Netherlands (and other European countries) local distribution network capacity falls short for the connection of newly built solar and wind parks. At the same time, distribution network capacity can hardly keep up with the fast development of new demand in urban areas, for example, in the Amsterdam metropolitan area where many data centres have been (and are being) established. It also remains to be seen if the current configuration of the power generation market can be maintained if, for instance, nuclear power plants turn out to be needed to achieve climate policy goals.

The reform of the natural gas sector was given shape in a similar way, with many actors, retail companies as well as distribution network providers, playing double roles in both the gas and electricity sector. While the national transmission network operators for gas and electricity are different entities, most regional distribution network operators provide both gas and electricity distribution capacity, and most energy service providers sell both gas and electricity. The gas sector is facing drastic change as natural gas is bound to be phased out for space heating purposes while industries are also looking for climate neutral alternatives to satisfy their high temperature heat demands. For the future, many municipalities have turned their hopes to heat networks, all-electric solutions involving heat pumps, and hydrogen-based systems, the latter to replace natural gas by hydrogen or to provide flexibility in interaction with the electricity infrastructure (energy storage and conversion). It is obvious that there is a strong dependency between the choice of energy supply system(s) and the necessary infrastructure layout for a specific location. Moreover, the technical and economic interdependency between electricity, gas and district heating systems is bound to increase, given the need to create supply and demand flexibility, which represents a radical break with the demand driven systems of the past.

Transaction Cost Economics in Heat Networks

To date, more than 90% of Dutch households still rely on natural gas for space heating. In the Netherlands, in 2018, the share of natural gas in the total final energy consumption amounted to 34.1%, which is far more than in any other EU Member State (averaged over the EU 27, the share of natural gas in the total final energy consumption was 20.8% in 2018) (EC, 2019c). Around 400,000 households are connected to local district heating networks. It is worth noting that most of the current networks are not supplied by renewable heat sources, but depend on natural gas-fuelled Combined Heat & Power (CHP) units. So far, private actors have shown little interest in establishing and operating sustainable heat networks. In reviewing the potential configurations shown in Fig. 1, it is evident that the ideal and less ideal market configurations do not fit with the transactions involved in investing in a district heating system and its long-term operation and exploitation. The necessary market is at best in a nascent stage. It will take a long time before it will have matured in terms of a sufficient number of competitive suppliers, sufficient diversity in supply and demand profiles, and costs. The return on the massive investment required will have to be realised over decades (40 years rather than 30), which is a period wrought with uncertainty given the technology dynamics in different types of heat sources (waste heat, geothermal heat, aquathermal heat, etc.), types of energy carriers (waste, biomass, green gas, hydrogen) and different scales of operation (from apartment buildings to neighbourhoods, districts, cities, regions etc.). Given the large differences in scale, complexity and flexibility of the different heating technology options, it is highly unlikely that a stable market will evolve soon.

Besides the technology dynamics, the heterogeneity of local heat supply and demand is a complicating factor. The choice of technology not only depends on the local availability of renewable heat sources, but also on the nature and quality of the local building stock. On both ends, different regions, different cities in the same region and even different neighbourhoods in the same city may be widely different. While the quality of the building stock defines the required temperature level of heat supply, it is highly uncertain if and how fast the building quality can be improved to the extent that low temperature heat supply would be adequate.

For many renewable heat sources, such as surface water, ground heat and deep geothermal heat, additional uncertainties are in store as climate change unfolds. The availability of underground heat and surface water may be affected by an increasing frequency and intensity of droughts and periods of rainfall. Geothermal heat extraction poses risks for contamination of groundwater aquifers that are now used for drinking water. These and other uncertainties require coordination with public actors and government bodies and imply an extent of risk exposure which private actors are not willing to accept.

For renewable or at least climate neutral heat supply to be accomplished in the future—if it is to be distributed through district heating networks—municipal governments are the authorities to take responsibility and to move first. The municipal government level fits best with the natural scale of heat networks, as dictated by the technology of heat distribution. It is evident that private actors cannot be expected to enthusiastically engage in such risky transactions, whether in the initial investment to bring a heat network into being, or in its long-term operation. If private actors are willing to take that risk, governments should be wary of the rewards these actors claim to cover their risk exposure. It is quite unlikely to be a good deal for the end-users that are, other than in the case of natural gas, captive users. With regard to natural gas services, consumers have a choice between different retailers. In the case of heating services, however, consumers are captive of the local service provider, which according to the current legal framework will be a vertically integrated private monopolist holding a long-term concession for the implementation and exploitation of the local heat network. That implies that extra caution is due for the protection of end-users, and that safeguards are needed in a long-term perspective. Such caution requires a high knowledge level with the municipal government, which may be lacking in many of the smaller cities. Even more stringent knowledge demands should be imposed on the municipal governments, if they themselves are to build and operate local heat networks, as public providers, which in fact is the optimal solution suggested by Williamson's theoretical framework.

Considering the knowledge position and operational experience of the energy distribution companies, including their deep insights in the energy needs and use patterns of their end-users, they seem to be the most appropriate actor to build and operate heat distribution networks. If and when hydrogen will be considered an appropriate substitute for natural gas in boilers and home-scale central heating systems, the established natural gas distribution networks may be used again. In the meantime, the established gas networks are being kept in mint condition for safety reasons. Even in districts where municipalities plan to phase out natural gas by 2030, planned maintenance and replacement works on the natural gas system cannot be relaxed or delayed. The natural gas networks can quite easily be adapted for the safe distribution of hydrogen, at relatively little cost in comparison with the construction and operation of heat distribution networks. Given the time horizon of 2030 that many municipal governments set to abandon natural gas, however, they fail to see hydrogen as a viable alternative, despite the obvious advantage of the availability of the network for its distribution. Indeed, to become a large-scale viable solution, a large enough production of green and blue hydrogen is required at an affordable price. It is questionable if this condition can be satisfied by 2030. Moreover, the process industry is most likely the primary candidate for hydrogen consumption, as its alternative options to generate high temperature heat and to change to low temperature electrochemical conversion routes are limited.

So far, according to the current energy acts in the Netherlands, heat (networks) should be provided in competition (for a concession to supply a certain area), and energy distribution companies are not eligible as heat network providers. Given the present institutional configuration, the projected outcome of the emerging heat 'market' is a patchwork of local heat networks with large differences in the quality and costs of service, depending on the local availability of sustainable heat sources, the topology of supply and demand, and the quality of the local building stock. It remains to be seen, however, if Dutch society is willing to accept significant cost differences for heat services between neighbourhoods, between cities and/or between regions. A public debate on the acceptability of such cost differences, parting from the principle of cost socialisation as traditionally followed in the provision of electricity and gas services, has yet to begin.

Emerging Value Tensions in Energy Infrastructure Governance

As we can see, the guidance provided by Williamson's framework for the institutional design of energy infrastructure (services) provision is wearing out as new societal priorities are coming to the fore. The focus of Williamson is primarily on creating an optimal contracting structure, from the perspective of economic efficiency. And indeed, the energy infrastructure restructuring of the 1990s and subsequent development of the sector was essentially driven by a focus on economic efficiency, through the creation of markets and competition for the electrons and gas molecules, and through regulation of the transport networks. Over time, this single-objective orientation has been replaced by a variety of objectives, as expressed and articulated by society. Today, first and foremost, CO_2 emissions are to be reduced, leading us to the selection of a collection of new technologies, which dictate different spatial patterns of production and transport. The proposed European Climate Law, however, which is intended to make the European ambition of net zero greenhouse gas emissions by 2050 a legally binding target, has yet to be adopted and come into force. While some Member States have embedded renewable energy and greenhouse gas emission reduction target in national legislation, most have so far been reluctant to take legally binding measures. In this respect, also the Netherlands' government was criticised for its lack of long-term political commitment to climate policy goals, and a corresponding lack of firm institutional safeguards (Faber et al., 2016).

In addition, the public acceptance of existing and new energy production facilities, transport and storage infrastructure has become much more difficult. Processes of planning and permitting have become highly complex and politicised, even among different levels of public authorities. This relates to societal values related with the protection of landscapes and nature areas, the preservation of urban structures, buildings and 'views', and also to the (expected) hindrance experienced by the people living near such structures. Local resistance adds to the long leadtime of infrastructure capacity expansion and new investment projects, once proposed investment projects have been approved by the regulatory authorities. The latter, as discussed previously, are mainly weighing the criterion of cost-effectiveness in their assessment. In the current perspective on assessing infrastructure investment projects, transport infrastructure is expected to follow capacity demand. Proactive investment in infrastructure capacity is not encouraged, which explains why new solar and wind parks in relatively remote areas may have to wait many years for capacity expansion of the local electricity distribution network.

In addition to a perceived lack of distributional justice in the allocation of financial benefits between land owners and homeowners on the one side, and tenants on the other side, such feelings of unease are also felt with regard to national subsidies for renewables leaking away from the country. More than three quarters of the large-scale solar parks in the Netherlands are owned by foreign investors. Of the total of Euro 1.1 bln in subsidy paid for the 33 largest solar parks in the Netherlands, almost Euro 890 mln leak away across the national border. Similarly, public discontent is emerging with the long-term contracting of large volumes of windpower by newly established large-scale data centres, owned and operated by large data multinationals. On the positive side, these contracts provide certainty of income to these wind parks, thus supporting their development. Yet, on the other side, the people living near the wind farms feel burdened by the environmental impact without benefitting, while the data-centres 'take away' the scarce sustainable power generated from them and from other domestic users, without creating employment opportunities in return. At the same time, many of these large-scale data centres are built in cherished polder landscapes, thus destroying their cultural-historic value. These sentiments of perceived injustice may turn into a serious issue, as domestic consumers are blamed for not being green enough, while they are confronted with higher tariffs and while increasingly large amounts of power are contracted away by large industrial users, like data centres and, in the future, the huge industrial producers of green hydrogen.

Moreover, from a relatively homogeneous utility provided public 'good' in the past, electricity and gas have been transformed into experience, identity or lifestyle goods and services, by which consumers, producers and prosumers exhibit and express their 'identity' and their values and convictions. Consequently, a rather divergent set of (instable) values is projected upon the societal function of energy supply. While the values of universal access and of affordability and availability of energy services remain uncontested, the value set contained in the overall value of acceptability of energy services is becoming over-crowded. Traditionally, acceptability was mainly concerned with health, safety and environmental concerns. While these values remain important, a new range of personal, cultural, ethical and social values is emerging in relation to the energy system, such as privacy and cyber security, historic and aesthetic value, fairness and justice, transparency and legitimacy of decision-making, and social inclusiveness. Conflicts arise between local values, such as landscape protection, and global values, like combating climate change, in solidarity with vulnerable people around the globe. The conflict is often presented as a false dilemma between the noble global cause on the one hand, and local jobs and affordability of energy services on the other hand. Such value conflicts can only be solved in political decision-making. In this respect we also have to realise that this emergent set of values differs between cultures and countries, leading to different priorities and to differences in the way each value is interpreted and made operational over time.

The question then arises how the coordination and governance of such a multi-value driven system should be shaped. The economic efficiency focus of Williamson's contracting scheme will evidently not suffice any more. All the more so, because from the relatively singular objective of a national (and even EU-wide) focus on efficiency, the orientation is shifting to a more locally specific shape of local energy systems, with local values and characteristics to be sustained. Logically, however, (almost) no local energy system can function stand-alone, all the way, all the time. This is all the more so, when we take into consideration the fact that the goods consumed in a specific area 'contain' large volumes of energy that are 'imported' from other areas. Hence, the question of above-local coordination and interconnection remains of the utmost importance.

Infrastructure and Inequality

According to the Dutch Energy Research Centre ECN (2017) 2.6 million low-income households in the Netherlands spend on average 9% of their net disposable income on energy, and the percentage of households that spends more than 10% of the household budget on energy costs has increased by 40% between 2006 and 2009. The threshold value of 10% of the household budget is often regarded as an indicator of energy poverty. We are only talking about the costs of energy services, which in turn are determined strongly by the nature and quality of housing: in that sense the entire built environment could be regarded as part of the infrastructure. Investments in home improvement or in sustainable energy call for resources which lower-income groups generally do not possess, money, knowledge, space and so-called 'acting ability', a term devised by the Netherlands Scientific Council for Government Policy (WRR, 2017). Meanwhile higher-income groups and homeowners are reaping the benefits of incentive schemes for investments in home insulation, sustainable energy and expensive electric mobility. This is worrying not only from an ethical perspective of justice. A distribution of costs and benefits that is perceived as unfair carries the risk of undermining public support for the substantial investments that need to be made for a sustainable energy supply and climate adaptation. We can learn so from Germany, where support for the *Energiewende* is eroding, firstly due to the perception of unfairness in the apportionment of costs and benefits and, secondly, due to the promised benefits in terms of jobs and CO_2 emission reductions failing to materialise (Andor et al., 2017).

The transformation from a carbon-based to a climate neutral economy comes with the risk of aggravating inequality in society. This is not only about the inequality arising along the traditional lines of income differences. The phase-out of natural gas as a heating fuel from the Dutch energy system puts great pressure on home improve-ment, which will hit especially private homeowners hard financially. Almost 90% of privately owned homes in the Netherlands predate 2005; almost two-thirds predate 1985 (CBS et al., 2020). Making these residences suitable for the energy transition requires major investments. The Netherlands Environmental Assessment Agency (PBL) estimates that the average private home, with energy label D, requires an investment of approximately € 35,000 to achieve energy neutrality (PBL, 2020). As the intended arrangement stands, the private owners must pay this amount; if necessary, facilitated by a long-term loan. For most private homeowners these investment costs are prohibitive. Except for investments in solar roof panels, they will not see their investment paid back in savings on their energy bill over decades. Perhaps these investment costs can also be factored into the monthly energy bill. Homeowners and occupants of older houses who cannot or will not invest in home improvement run the risk of being stuck with unsaleable property. If not remedied by the government, the energy transition may thus imply a massive redistribution of wealth among the population. Tenants of private homeowners also constitute a vulnerable group; their energy bill depends on investments by the property owners. For the less educated homeowners and occupants it will by no means be easy to make a sensible and well-informed choice from the different options for heating supply without natural gas in their district.

Other forms of inequality can be caused by the digitalisation of infrastructure-related services. For less educated, semi-literate or digitally illiterate citizens it is hard enough as it is to assess the different offers of energy suppliers. Things will become even more complicated when they are also expected to take part in active demand response programmes. Part of the problem is that it will be difficult for them to make an informed choice between the schemes offered by different retailers. The real problem though arises from the fact that many of these citizens live in energy poverty, in energy efficient dwellings. As their energy use is limited to satisfying basic needs of comfort, implementation of automated demand response schemes can hardly contribute to reducing their energy bill. Their situation will only be improved by structural home improvements, requiring financial means they do not have.

Whereas the universal access to the supply of electricity and heating is in any case still regulated by law, that principle is not applied to the transport infrastructure nor for public transport. Although it is a responsibility of regional and local authorities to ensure that transport options in their regions and districts are sufficiently accessible to those who do not have or cannot afford a car of their own, there is no statutory basis or a hard norm for minimum accessibility e.g., in terms of maximum distance to a public transport boarding point, minimum frequency of the service or the tariff structure. In addition, for the time being there is no accepted indicator for transport poverty, expressed in the share of the necessary transport costs in the spending of the household budget. Although in comparison with other countries in Europe, and certainly in comparison with the USA, the Netherlands is amply furnished with good and affordable mobility options, in which bicycles are just one important element, transport poverty does exist in the Netherlands as well (Bastiaanssen et al., 2013). Transport poverty, like energy poverty, is a multi-faceted problem, which is found all over the world in developing as well as developed economies (Lucas et al., 2016a, b).

In the Netherlands, lack of access to or problems with the affordability of public transport affects inter alia the older population in regions with shrinking population. As other essential facilities such as medical services, shops, bank branches, ATMs become scarcer there, the transport demand of the carless population changes and they tend to face ever more transport poverty. Situations of transport poverty also occur in urban districts with a low socioeconomic status. Rotterdam-South is an example in kind. Admittedly, this district is adequately connected to the city centre via bridges, the Maas tunnel and the metro, but most of those connections are of importance mainly for car owners. For the majority of inhabitants of the Rotterdam-South district a car is an unattainable ideal, whereas the public transport network provides few opportunities to reach work sites in the region, certainly outside the city centre and outside normal office hours, aside from the costs of the metro or bus fare (Bastiaanssen et al. 2013). Transport poverty, whether in terms of affordability or from lack of physical accessibility, thus hits people hard in their development opportunities. It can make work and proper schooling inaccessible, as well as visits to friends and relatives (Bastiaanssen et al., 2020).

It is a fact that for investments decisions in transport infrastructure the interests of entrepreneurs and car owners have a relatively big weight, because their willingness to pay for savings in travel time is relatively strong. For financially weak groups in society investments in high-quality public transport facilities are hardly justifiable, if that consideration is placed primarily in an economic perspective, as it is in the SCBA methodology.

New Challenges for Infrastructure Policy

Inequality as such is not unacceptable and cannot always be avoided. Our society accepts a considerable degree of inequality. However, if inequality concerns the accessibility and affordability of essential infrastructural services, which everybody needs in order to function in society, this affects ethical values of justice. In that sense, the transition from natural gas to alternative forms of heating supply at urban, district or building level, as in the Netherlands, is more than just a technological transition. It is also a transition from a public amenity at national level, with socialisation of the costs, to an individual system or to a new collective service at the scale of the street, the district or the city, in which the solidarity principle of the natural gas infrastructure is abandoned. It is not self-evident that this transition will lead to strengthening of the social cohesion at district level, if there is no policy in place to guard against great inequality in the quality and affordability of heat services and level of comfort. There is most certainly a risk of new divides arising in the population. In many countries, such divides manifest themselves literally in the walls and fences around gated communities.

Social cohesion is not the only issue in this context, though. It concerns individual development opportunities for every member of society. It goes without saying that those opportunities are not determined exclusively, but definitely to a significant extent, by the access to and affordability of infrastructure services. If work sites cannot be reached by a part of the population, society is not only compromising the interests of those citizens, but also the interests of society itself, by underutilising its supply of productive labour. This also goes for access to education, for access to healthcare and cultural facilities. When part of the population can only maintain their social contacts in the digital domain, there is a real risk that people who cannot travel physically may become socially isolated. That, too, does not affect only those citizens. It affects society as a whole, which will lose quality and cohesion.

The Covid-19 pandemic has exacerbated inequality in society. It has disproportionally hit people living in crowded households and in homes without outdoor space, and people without private means of transport. Children in households without access to the internet, whether by lack of a connection or by lack of access devices, have been deprived of schooling. Wherever physical infrastructure and infrastructural services are required for citizens in order to function as full members of society, public policy attention is needed. In this respect a statutory basis of standards for accessibility and affordability of fast internet and public transport options may be considered, or a more active stimulation of citizens' initiatives to close gaps in the public transport network and timetables. Now such bottom-up initiatives sometimes fail on account of alleged competition with the minimal public transport that is still being offered. Moreover, a perspective of justice can be made operational in indicators which can be incorporated into the SCBAs, for instance by means of standards for transport sufficiency (ref. Lucas et al., 2016a, b; van Wee, 2012). In the present practice of SCBA, however, such social values are not yet included.

An insidious shift towards more inequality in society that is inadvertently encouraged by today's 'infrastructure policy' is an important reason for examining that policy in more detail. Here we put infrastructure policy in inverted commas, because it does not only concern infrastructure policy in the classical meaning, like the development of roads, railway lines and gas networks, or inviting tenders for public transport concessions. It also concerns policy with a big impact on the future development of those classical infrastructure systems, especially climate policy. That policy does not only have far-reaching consequences for the technical development of installations and networks which will in the future fulfil our needs for energy and mobility. It will also affect us profoundly in our behaviour as consumers, citizens, entrepreneurs and employees. Authors like (Shove, 2003; Shove et al., 2012), Van Vliet et al., (2005) and Overbeeke, (2001) show that infrastructure services are strong determinants for our everyday social routines and practices. Social norms for e.g., personal and household hygiene, for comfort and social interaction changed drastically in the course of the previous century under the influence of infrastructure provisions for drinking water and energy, waste and wastewater discharge, telecommunication and internet (social media). The possibility to refrigerate and deep-freeze food at home had a great impact on the food supply, via the supply and the location of regional supermarkets, which have largely taken over the role of neighbourhood shops, bakers, butchers, greengrocers and fishmongers. The access to and the reach of transport options in the form of public transport or roads and motorways, does not only impact decisions about housing and work sites, but also the recreation and the sociocultural behaviour of citizens (Raspe, 2012; Steg & Vlek, 2009; Teulings et al., 2017; Van der Knaap, 2002; Van Wee et al., 2013). Infrastructure affects all of us profoundly in our daily lives.

The presence and nature of infrastructure provisions are not only strong determinants for the use of energy and water, the mobility of people and goods and other aspects of social and economic routines, but also for the possibilities we have to develop our capabilities (Nussbaum, 2001; Sen, 2010). This is true both for individual citizens and for groups in society, for districts and for regions. It means that big changes in the supply of essential services must be assessed not only for their direct effects—for instance savings in travel time in connection with the construction of new roads or reductions in CO_2 emissions in case of adjustments to the energy supply - but also for their indirect effects on society.

Towards a New Public Debate

An important conclusion of this chapter is that **not every citizen is affected to the same extent by the major changes that occur and will occur in the provision of infrastructure services**: whereas new development opportunities arise for some, others will be deprived of them. This influences an essential aspect of the role of infrastructure as a binding factor in our society. For a long time, infrastructure could be regarded as the fabric of society, on the one hand because infrastructure provision literally connected everybody, on the other hand because infrastructure equipped every member of society with more or less equal development opportunities. Postage stamp tariffs for private end users explicitly expressed the principle of solidarity in the world of infrastructure services. The ongoing technological and administrative decentralisation of infrastructure development, in combination with the introduction of market forces, may erode that binding role of infrastructure as the fabric of society. The question is whether the decision makers about infrastructure are sufficiently aware of this risk.

Inequality between citizens is a given. It is for good reason that we cherish our individuality. In this chapter, the focus is on inequality created by the organisation of infrastructure systems and changes therein, whether induced by new technologies or changing societal preferences. We need to be concerned about the bandwidth of the additional inequality that may be, albeit unintentionally, caused thereby and the balance in time. Public investments in innovative infrastructure services, which initially are not accessible to everybody and everywhere, are justified if such investments contribute to innovation, economic growth and new employment opportunities or to a more affordable or better functioning system that will in the longer term benefit all members of society. If, however, some regions or groups in society are excluded from those benefits, also in the longer term, there is every reason to recalibrate infrastructure policy.

This chapter is a plea to fundamentally reconsider the landscape of infrastructure provisions. On the one hand, a large part of the infrastructure is ageing; on the other hand, we are confronted with the new challenges of urbanisation, digitalisation, climate change, and so forth. Especially the energy transition will deeply influence all realms of the economy and society, as all economic activity, in each and every sector, depends on energy. Consequently, nearly all infrastructure systems are facing great investment issues. These are complicated issues, which we cannot solve one by one in isolation, given the interdependencies in space and functionality between different infrastructure systems. Accounting for these multi-scale and cross-sector interdependencies forms part of the challenge we face in infrastructure policy making and governance. Some countries have already added a cross-sector body to the siloed infrastructure governance structure in order to provide for a cross-sector assessment framework. As an example, we refer to the National Infrastructure Commission (NIC) in the United Kingdom, which was established as an executive agency in the Treasury to advise the UK government on all issues pertaining to long-term infrastructure challenges. Building on the National Infrastructure Assessment work of the NIC, the UK government presented its cross-sector National Infrastructure Strategy in November 2020 (NIC, 2018, 2020; UK Gov, 2020).

There is an accumulation of revolutions unfolding in the technological and institutional organisation of infrastructure systems. A cohesive overview of and insight into the consequences of this for society in the longer term are missing, among other things because the field is highly fragmented: in addition to citizens developing bottom-up initiatives, there is a large number of public and private actors involved in each of the infrastructure domains, all active at different geographical and administrative levels. Each infrastructure sector, or each infrastructure network, has its own legislative and regulatory framework, its own supervisory body, its own policy silo, at the national level and at the European level. Within those frameworks, actors operate according to the statutory allocation of roles and appurtenant mandates and responsibilities. This implies that the current institutional organisation of infrastructure policy generates few incentives for coherence across the boundaries of infrastructure domains. This is despite the fact that an intervention in one infrastructure system also affects other systems in the infrastructure system-of-systems, and that the consequences will affect every citizen and the relations between citizens.

What we also know for certain is that the infrastructure development choices we make today will have a far-reaching impact on the future. This creates a responsibility to future generations. It is because of that very responsibility that we should not rush into thinking in technical solutions, but should ask ourselves first: How do we want to live together? We are all inseparably connected with infrastructure, at least as end-users, and via infrastructure we are interconnected directly and indirectly. Infrastructure binds us as a society. In this context, we may cite the UK Prime Minister, Boris Johnson, in his foreword to the National Infrastructure Strategy, where he re-iterates his promise *".... to unite our country by physically and literally renewing the ties that bind us together"* (UK Gov, 2020). Infrastructure makes it possible for us to live together and engage in various communities, in formal and informal relations of families, the neighbourhood, the city, in other social networks and all kinds of organised forms of activity. Infrastructure is not something outside society; it forms an essential part of it, not only as the connective fabric of the economy, but of society as a whole.

In that perspective, it makes sense to see ourselves as part of the infrastructure system. That perspective is a break with the past. Whereas infrastructure was traditionally regarded as a collection of technical components, managed by the government, we now see a system that is directed by a constellation of public and private actors. The image fitting with the current organisation of infrastructure is that of a sociotechnical system, a system that is determined as much by social actors and institutions as by technology. Apart from network managers, suppliers, traders, market operators, policymakers, regulators, and so forth, as users we also belong to the network of social actors that co-evolves with the physical networks. Moreover, our role in the infrastructure system is no longer that of passive end users. Now that we are being asked increasingly to present ourselves as active, flexible end users and even as service providers in the market, the role we play as actors in the system is acquiring a deeper meaning. For all the actors in the system their behaviour is directed by institutions. These include technical and operational standards, types of ownership and contractual arrangements, but also policy choices for regulation and market forces, as well as the standards and values of our society.

That knowledge enables us to enrich and deepen the infrastructure debate substantially. It is of great public interest that we should become aware of the values that are embedded in the infrastructure that serves us, and of the values which we would want to anchor in new infrastructure being developed today. Infrastructure development is no longer a matter of technocratic or unilateral economic policy; it is a matter of social debate and political choices.

In the present debate ever terser questions about social justice and inclusivity emerge. What are the dominant values of the society we want to be? And what do they mean for the governance of the infrastructure? How can those social values be expressed in policy documents, draft plans and implementation projects of infrastructure? In this context, we can also point to the European Pillar of Social Rights, which was proclaimed in Gothenburg by the European Parliament, the European Council and the European Commission on 17 November 2017 (European Commission, 2017). This Pillar mentions right of access to essential services like transport, energy, digital communication, water, sanitation and financial services as a social right for all European citizens.

We argue that the infrastructure debate should urgently pay more attention to the social values at stake in the big changes that are unfolding in society and in the transformation ahead, especially with regard to the energy transition. Without such explicit attention, society is at risk of an intensification of inequalities and of the emergence of new inequalities in the accessibility and affordability of infrastructure services, which are undesirable from a viewpoint of social justice. **The fundamental role of infrastructure as *fabric of society* appears to be a blind spot in reflections on infrastructure and largely unexplored territory in infrastructure policy and governance**. With this contribution we intend to spark a richer, strategic debate about the role that infrastructure can and should play in the future development of society.

After all, infrastructure is not an end in itself, but a means to enabling and shaping the society that we want to be. Society does not care about infrastructure, but about accessibility, connectivity, mobility, comfort, health, social connectedness, development opportunities, and so on. This is true for young and old alike, for rich and poor, healthy and disabled, computer literate and illiterate, highly and poorly qualified, in cities and rural areas. Infrastructure is essential to enable all citizens to take part in society and, with the means and possibilities at their disposal, to create new social and economic value for that society.

References

Andor, M. A., Frondel, M., & Vance, C. (2017). Germany's Energiewende: a tale of increasing costs and decreasing willingness-to-pay, *IAEE Energy Forum*, 4th quarter 2017: 15–18.

Aschauer, D. A. (1989). Is public expenditure productive? *Journal of Monetary Economics, 23*(2), 177–200

Barabási, A.-L., & Albert, R. (1999). Emergence of Scaling in Random Networks. *Science, 286*, 509–512

Bastiaanssen, J., Henk, D., & Karel, M. (2013). Vervoersarmoede. Sociale uitsluiting door gebrek aan vervoersmogelijkheden. *Geografie*, October 2013: 6–10.

Bastiaanssen, J., Johnson, D., & Lucas, K. (2020). Does transport help people to gain employment? A systematic review and meta-analysis of the empirical evidence. *Transport Reviews, 40*(5), 607–628. https://doi.org/10.1080/01441647.2020.1747569

Batty, M. (2008). *Cities as complex systems: scaling, interactions, networks, dynamics and urban morphologies*. UCL Working Papers Series, Paper 131, February 2008.

Bettencourt, L. M. A., José, L., Dirk, H., Christian, K., & Geoffrey, B. W. (2007). Growth, innovation, scaling, and the pace of life in cities. *PNAS* April 24, 2007, *104*(17), 7301–7306. https://doi.org/10.1073/pnas.0610172104

British Medical Journal (2007) *BMJ* 2007, 334 https://doi.org/https://doi.org/10.1136/bmj.39097.611806.DB (Published January 18, 2007)

Carlsson, R., Otto, A., & Hall, J. W. (2013). The role of infrastructure in macroeconomic growth theories. *Civil Engineering and Environmental Systems, 30*(3–4), 263–273. https://doi.org/10.1080/10286608.2013.866107

CBS. (2019). *Value added of infrastructure 1995–2016, statistics Netherlands*. CBS, August 2019. See https://www.cbs.nl/en-gb/custom/2019/35/value-added-of-infrastructure-1995-2016

CBS, PBL, RIVM, WUR. (2020). *Woningvoorraad naar bouwjaar en woningtype, 2019* (indicator 2166, versie 04 , 20 oktober 2020). www.clo.nl. Centraal Bureau voor de Statistiek (CBS), Den Haag; PBL Planbureau voor de Leefomgeving, Den Haag; RIVM Rijksinstituut voor Volksgezondheid en Milieu, Bilthoven; en Wageningen University and Research, Wageningen.Compendium voor de Leefomgeving (2019). https://www.clo.nl/indicatoren/nl2166-woningvoorraad-naar-bouwjaar-en-woningtype

CPB. (2018a). How large are road traffic externalities in the city? The highway tunneling in Maastricht, the Netherlands. Centraal Planbureau, 22 May 2018. See: https://www.cpb.nl/en/node/159454#publicaties

CPB. (2018b). *How large are road traffic externalities in the city? The highway tunneling in Maastricht, the Netherlands*, CPB Discussion paper 379, 25 April 2018. Centraal Planbureau.

de Bakker, M., Lagendijk, A., & Wiering, M. (2020). Cooperatives, incumbency, or market hybridity: New alliances in the Dutch energy provision. *Energy Research & Social Science, 61*, 101345

ECN. (2017). *Rapportage Energiearmoede. Effectieve interventies om energie efficiëntie te vergroten en energiearmoede te verlagen*, ECN Beleidsstudies. Energieonderzoek Centrum Nederland

EC. (2013). *European system of accounts ESA 2010, Eurostat*. European Commission. ISBN 978-92-79-31242-7. See https://ec.europa.eu/eurostat/documents/3859598/5925693/KS-02-13-269-EN.PDF/44cd9d01-bc64-40e5-bd40-d17df0c69334

EC. (2017). *European pillar of social rights*. European Commission, 16 November 2017. ISBN 978-92-79-74092-3. https://doi.org/10.2792/95934

EC. (2018). *Digital economy and society index report 2018*. European Commission. See: https://ec.europa.eu/information_society/newsroom/image/document/2018-20/1_desi_rep ort_connectivity_DFB52691-EF07-642E-28344441CE0FCBD1_52245.pdf

EC. (2019a) *Brussels: European commission*, 11 December 2019. See: https://ec.europa.eu/com mission/presscorner/detail/en/ip_19_6691

EC. (2019b). *Press remarks by President Von der Leyen on the occasion of the adoption of the European Green Deal*. European Commission, 11 December 2019. See https://ec.europa.eu/com mission/presscorner/detail/en/speech_19_6749

EC. (2019c). *Renewable energy statistics*. European Commission (2019). https://ec.europa.eu/eur ostat/statistics-explained/index.php/Renewable_energy_statistics

EC. (2020a). *Executive vice president timmermans' remarks at the conference on the first European Climate Law*. European Commission, January 28, 2020. See https://ec.europa.eu/commission/ presscorner/detail/en/SPEECH_20_144.

EC. (2020b). *Digital society and connectivity index 2020—connectivity*. European Commission. See: https://ec.europa.eu/digital-single-market/en/broadband-connectivity

Faber, A., de Goede, P. J. M., & Weijnen, M. P. C. (2016). *Long-term commitment for national climate policy in the Netherlands, WRR-Policy Brief no. 5*. Netherlands Scientific Council for Government Policy (WRR)

Frischmann, B. M. (2012). *Infrastructure—The social value of shared resources*. Oxford University Press. ISBN: 9780199895656

HIER (2020). *Lokale energiemonitor,* Hier opgewekt. See: https://www.hieropgewekt.nl/uploads/ inline/Lokale%20Energiemonitor%202019_DEF_feb2020_2.pdf, consulted 17–01–2021.

Hufen, J. A. M., & Koppenjan, J. F. M. (2015). Local renewable energy cooperatives: Revolution in disguise? *Energy Sustainable Social, 5*, 18. https://doi.org/10.1186/s13705-015-0046-8

Knaap, G. A. van der. (2002). *Stedelijke Bewegingsruimte, over veranderingen in stad en land*. WRR Voorstudies en achtergronden V113. Sdu uitgevers.

Lucas, K., Mattioli, G., Verlinghieri, E., & Guzman, A. (2016a). Transport poverty and its adverse social consequences. In: *Proceedings of the institution of civil engineers—Transport* (vol. 169(6), pp. 353–365). Published Online: November 18, 2016. https://doi.org/10.1680/jtran.15.00073

Lucas, K., Wee, B. Van, & Maat, K. (2016b) A method to evaluate equitable accessibility: combining ethical theories and accessibility-based approaches. *Transportation, 43*, 473–490. https://link.spr inger.com/article/https://doi.org/10.1007/s11116-015-9585-2

Mouter, N., Annema, J. A., & van Wee, B. (2013). Attitudes towards the role of Cost-Benefit Analysis in the decision-making process for spatial-infrastructure projects: A Dutch case study. *Transportation Research Part a: Policy and Practice, 58*, 1–14

Munnell, A. H. (1992). Policy Watch: Infrastructure Investment and Economic Growth'. *Journal of Economic Perspectives, 6*(4), 189–198

NIC. (2018). National Infrastructure Assessment: an assessment of the United Kingdom's infrastructure needs up to 2050: UK National Infrastructure Commission, July 10, 2018. See: https://nic.org.uk/studies-reports/national-infrastructure-assessment/#:~:text=An%20asse ssment%20of%20the%20United%20Kingdom's%20infrastructure%20needs%20up%20to% 202050.&text=The%20Assessment%20analyses%20the%20UK's,identified%20needs%20s hould%20be%20met.

NIC. (2020). System analysis of interdependent network vulnerabilities. Report by ITRC on the vulnerabilities of multi-sector infrastructure networks: UK National Infrastructure Commission, May 28, 2020. See: https://nic.org.uk/studies-reports/resilience/system-analysis-of-interdependent-network-vulnerabilities/.

Nussbaum, M. C. (2001). *Upheavals of thought: The intelligence of emotions.* Cambridge University Press.

ONS. (2018). Developing new statistics of infrastructure: August 2018, Office of National Statistics, United Kingdom, August 21, 2018. See: https://www.ons.gov.uk/releases/developingnewstatisti csofinfrastructureaugust2018.

PBL. (2020). *Woonlastenneutraal koopwoningen verduurzamen: verkenning van de effecten van beleids- en financieringsinstrumenten.* Planbureau voor de Leefomgeving, 2020. PBL-publicatienummer: 4152. See: https://www.pbl.nl/sites/default/files/downloads/pbl-2020-woonla stenneutraal-koopwoningen-verduurzamen-4152.pdf.

Raspe, O. (2012). *De economie van de stad in de mondiale concurrentie,* in *Essays Toekomst van de Stad* (pp. 20–24), Raad voor de leefomgeving en infrastructuur.

Romijn, G., & Renes, G. (2013). *Algemene leidraad voor maatschappelijke kosten-batenanalyse.* Centraal Planbureau (CPB).

Sen, A. (2010). *The Idea of Justice.* Penguin Books.

Shove, E. (2003). *Comfort, cleanliness and convenience: the social organization of normality.* Bloomsbury Academic.

Shove, E., Pantzar, M., & Watson, M. (2012). *The dynamics of social practice: everyday life and how it changes.* Sage Publications.

Statistics Canada. (2018). Infrastructure economic account. See: https://www150.statcan.gc.ca/n1/ pub/13-607-x/2016001/1362-eng.htm.

Steg, L., & Vlek, C. (2009). Encouraging pro-environmental behaviour: An integrative review and research agenda. *Journal of Environmental Psychology, 29*(3), 309–317

Straub, S. (2008). *Infrastructure and development : A critical appraisal of the macro level literature,* Policy Research Working Paper; No. 4590. World Bank. See: https://openknowledge.worldbank. org/handle/10986/6517 License: CC BY 3.0 IGO.

Thacker, S., Adshead, D., Fay, M., et al. (2019). Infrastructure for sustainable development. *Nature Sustainability, 2*(4), 324–331

Teulings, C. N., Ossokina, I. V., the Groot, H. L. F. (2017). Land use, worker heterogeneity and welfare benefits of public goods. *Journal of Urban Economics, 103,* 67–82.

Tomer, A. (2018a). *A "people first" perspective on infrastructure: Delivering access.* Brookings Institution. See: https://www.brookings.edu/blog/the-avenue/2018/05/08/a-people-first-per spective-on-infrastructure-delivering-access/

Tomer, A. (2018b). *Can people afford American infrastructure?* Brookings Institution. See: https:// www.brookings.edu/blog/the-avenue/2018/05/09/can-people-afford-american-infrastructure/

UK Gov. (2020). *National infrastructure strategy: Fairer, faster, greener.* Presented to Parliament by the Chancellor of the Exchequer, November 2020: UK Government. See: https://assets.publishing.service.gov.uk/government/uploads/system/uploads/attachment_d ata/file/938049/NIS_final_web_single_page.pdf

UN. (2009). *European communities, international monetary fund, organisation for economic co-operation and development, United Nations and World Bank. 2008 system of national accounts.* United Nations. ISBN 978-92-1-161522-7. See: https://unstats.un.org/unsd/nationala ccount/docs/sna2008.pdf

Verkade, N., & Höffken, J. (2019). Collective energy practices: A practice-based approach to civic energy communities and the energy system. *Sustainability, 11*(11), 3230

van Overbeeke, P. (2001). *Kachels, geisers en fornuizen: Keuzeprocessen en energieverbruik in Nederlandse huishoudens 1920–1975.* Uitgeverij Verloren.

van Vliet, B., Shove, E., & Chappells, H. (2005). *Infrastructures of consumption: Environmental innovation in the utility industries.* Earthscan.

van Wee, B. (2012). How suitable is CBA for the ex-ante evaluation of transport projects and policies? A discussion from the perspective of ethics. *Transport Policy, 19*(1), 1–17

van Wee, B., Annema, J. A., & Banister, D. (2013). *The transport system and transport policy: An introduction.* Edward Elgar Publishing Ltd.

WEF. (2012). *The Europe 2020 competitiveness report: Building a more competitive Europe.* World Economic Forum.

WEF. (2017). Global competitiveness report 2017–2018. World Economic Forum, September 26, 2017.

WEF. (2019). *Global Competitiveness Report 2019: How to end a lost decade of productivity growth.* World Economic Forum, October 8, 2019.

Williamson, O. E. (1998). Transaction cost economics: How it works; where it is headed. *De Economist, 146*(1), 23–58

WRR. (2017). *Weten is nog geen doen. Een realistisch perspectief op redzaamheid.* WRR-rapport nr. 97. Wetenschappelijke Raad voor het Regeringsbeleid.

Woud, A. van der. (1987). *Het lege land. De ruimtelijke orde van Nederland 1798–1848.* Meulenhoff Informatief.

Woud, A. van der. (2020). *Het landschap, de mensen. Nederland 1850–1940.* Promotheus.

Perspectives on Justice in the Future Energy System: A Dutch Treat

Aad Correljé

Abstract The (un)affordability, the (un)reliability and the (un)sustainability of our energy supply are increasingly associated with the phenomenon of energy justice. This concerns the way in which different groups of citizens and businesses experience the benefits and burdens of the energy transition. We explore how the concept of energy justice may support a just transition. Firstly, we address the socio-political embedding of the energy sector and policy-making. Then we explain how the concept of energy justice is defined and operationalized, in respect of policy making and implementation. Thereupon we apply the concept of energy justice to the current Dutch energy debate, addressing the reduction of natural gas production to diminish the number and strength of earthquakes in Groningen, and the longer-term policy objectives of the energy transition. It addresses the radical changes in energy use and supply and the consequent wide variety in direct and indirect consequences for citizens and businesses, depending on their specific circumstances. The notion of energy justice is discussed as a feature in local, national and EU policy making and implementation, and as a claim of social actors, communities and individuals. The suggestion that justice issues can be identified and solved at these levels, is too simple. It is important to consider the layout and nature of the socio-technical energy system and its functioning. It is concluded that the concept of justice may help researchers to identify the relevant values and value conflicts in the energy transition. This can help policymakers to make informed choices.

Introduction

It can be argued that the traditional trinity of energy policy objectives—affordable, reliable and sustainable—is facing competition from a fourth candidate target: energy justice. This looks like an interesting proposition. Of course, we want energy supply, as the driving force of our society as a whole, as well as of the social functioning

A. Correljé (✉)
Faculty of Technology, Policy and Management (TBM), Delft University of Technology, Delft, The Netherlands
e-mail: A.F.Correlje@tudelft.nl

© The Author(s) 2021
M. P. C. Weijnen et al. (eds.), *Shaping an Inclusive Energy Transition*,
https://doi.org/10.1007/978-3-030-74586-8_3

of the individuals being part of it, to fulfil the three traditional objectives. Yet, it is equally important to have an eye for the ethics of a just distribution of the benefits and burdens that are associated with the provision of energy to society.

In the public debate, the (un)affordability, the (un)reliability and the (un)sustainability of our energy supply, now and in the future, are increasingly associated with the phenomenon of justice. This often concerns the way in which different groups of citizens and businesses, to a greater or lesser extent, experience the benefits and burdens of the current system of energy supply. Yet, it is particularly in the context of the energy transition and the drastic changes foreseen that energy justice is brought in as a major policy objective. Indeed, citizens will experience such changes, for example, in their role as homeowner or tenant, as residents of a specific municipality, as a member of a local community living near a particular energy installation, as a traveller, as a consumer, as a saver or shareholder, as a taxpayer, as an employee, and in their social awareness, or—very personally— in their mental well-being. Businesses will also experience the transition, either as consumers or producers and suppliers of energy. It will change their operational processes and investment decisions and their purchasing and selling activities in markets. It will also influence their relations with governments when it comes to the awarding of permits and licenses, the impact of taxes and subsidies, and ultimately in their business results.

The direct consequences of the energy transition can be positive for citizens, in the sense that their living environment, housing, employment and their opportunities for transport improve. For some businesses new activities and opportunities will arise. Others will suffer a decline. Moreover, there are the generic advantages of curbing the greenhouse effect and of other improvements to the living environment and nature. These advantages will be felt more or less strongly in different places. On the other hand, there are the negative consequences, in terms of the financial and social costs of introducing new technologies and the scrapping of old, businesses and regions going into decline, adaptation to new patterns, routines and practices, and also 'new' damages to the quality of living and living environment (Kooger et al., 2017; SER, 2018).

As these advantages and disadvantages will be spread in an unequal way among (groups of) citizens, businesses and localities, there clearly is a distributional issue at stake. The fact that this may arouse social discontent about the distribution of benefits and burdens is seen as a functional impairment of societal support and creating resistances against the transition (SER, 2017: 12). Moreover, also an important ethical issue arises here. This particularly concerns inequality, firstly, in the degree to which citizens will experience the direct effects of the transition and, secondly, in the way in which they are socially and economically able and capable to adjust their daily life and practices, to give a positive interpretation to the necessary changes (See Kooger et al., 2017).

In this paper we want to explore what the contribution of the concept of energy justice can be to a (just) transition and what new insights this may generate. For some time now, the notion of Energy Justice is put forward as a conceptual framework. It should enable us to determine whether certain developments in the energy system can be judged as "just" or ethically justifiable, or not. The aim of energy justice as a scientific approach is "to provide all individuals, across all areas, with safe, affordable and sustainable energy" (McCauley et al., 2013).

The question thus arises what 'just' may mean in the context of the provision of energy. Inspired by the doctrines of environmental and climate justice, McCauley et al. (2013) suggest three basic forms, or core tenets, of justice: (1) *distributive justice*, questioning how the benefits and burdens of energy supply and energy use are (spatially) divided among groups of people; (2) *procedural justice*, questioning how decision-making processes provide access to and participation to particular social groups and in which way; and (3) *justice through recognition*, defined as the need to recognize the dignity and rights of all individuals and the need for them to be included and therefore avoid the conditions of deprivation (such as that of fuel poverty). This addresses the way in which social, cultural, locational and other aspects structurally influence the exposure of groups to benefits and burdens and their capacity to deal with them. Recognition is then a precondition for trust and involvement and providing compensation.

Sovacool and Dworkin (2015) present energy justice as a conceptual approach. First, to relate various questions of justice to the provision and use of energy. Secondly, as an analytical approach for researchers to identify and operationalize the different values that play a role in the energy system. And thirdly, as a deliberative instrument for policy makers enabling them to arrive at more informed choices and policies. Jenkins (2018: 120) adds that energy justice offers a clearer focus on energy systems than the broader notions of environmental and climate justice because it develops an explicit energy focused methodology.

In the first section we will address the socio-political embedding of the energy sector and policy-making. Of importance is its evolution from a public service driven utility system in the past, to a more market-oriented—yet publicly coordinated— service focusing on efficiency, and more recently towards an energy system that is primarily driven by sustainability goals, in terms of a reduction of CO_2 emissions. We argue that an energy transition in a liberalized energy sector requires explicit attention for issues of energy justice. Indeed, there will be a great diversity in the way different (groups of) citizens and businesses will experience the consequences of the transition. This will be highly dependent on their specific social and economic circumstances, creating large contrasts in their possibilities to anticipate and to respond to the changing conditions under which energy will be supplied.

This motivates a more thorough examination of the concept of energy justice. Section "Public Values and Energy Supply" explains how the concept of energy justice is defined and operationalized in the academic literature, and provides possible connections with the practice of policy making and implementation.

Subsequently, in Section "Energy Justice", we will examine how the concept of energy justice may be applied in unravelling the current Dutch energy debate. In this debate, two policy objectives are paramount; firstly, on short notice, there is the need to reduce the supply of natural gas from the huge Groningen field, in order to diminish the number and strength of the earthquakes caused by the ongoing depletion of this field. Secondly, there is the longer term policy objective of the energy transition, requiring policy measures to be taken in the very near term, as well. In this context, we will discuss the notion of energy justice as an argument in national and local policy making, as a feature of the policy implementation at various levels (European Union, Netherlands, municipalities), and as a claim of social actors, communities and individuals. Finally, we will briefly address the question of how the concept of justice can contribute to a socially responsible policy for the energy transition (Section "Values in the Energy Transition").

Public Values and Energy Supply

The social importance of energy is nothing new. Energy supply systems and energy use always have had a strong influence on the social and economic functioning of societies, the people and their activities, and vice versa. This applies both to the availability of energy resources, as well as to the governance of the system of energy provision, within the prevailing socio-political context (see Goudsblom, 2001). The functioning of energy supply systems and their impact on society has attracted attention already for centuries; the impact of the peat dredging on water safety in eighteenth century Holland being a case in point (Rooijendijk, 2009). The notion of energy supply as a utility service arose in the beginning of the twentieth century. In Europe, the public values attached to the expansion of energy supply led to the establishment of public gas and power utilities, controlled by local and national authorities (Milward, 2005). Since the 1980s, we have seen a shift from public utility to market coordination, in which government only intervenes when and where necessary. Energy was transformed from a utility into a private commodity (De Jong et al., 2005).

Energy Supply as a Utility Sector

Over time, the pursuit of an affordable and reliable, and later also a sustainable, energy supply, has been given shape in different ways depending on the era. This has regularly led to debates and political discussions about the organization and the instrumentation of the supply of energy. Examples of these debates are the conversion of private gas and electricity companies into public, municipal, utilities in the first decades of the twentieth century; the large-scale introduction of natural gas in the 1960s; and the liberal restructuring of the energy sector during the 1990s (Hesselmans & Verbong, 2000; Hesselmans et al., 2000a, b, Correljé & Verbong, 2004). Obviously, such debates concerned distributive issues relating to tariffs and taxation and access to services and facilities. Later on, they also involved safety and environmental issues; like the broad public debate around nuclear energy in the early 1980s, acid rain, unleaded gasoline and the debate about responsible gas exploitation in the Waddensea nature reserve and the issue of compensation, at the beginning of this century. A variety of energy-related values were defined, articulated and placed on the political agenda, as an issue of public interest in evolving societal and political discussions (De Jong et al., 2005; van der Linde 2008; Groenewegen and Correljé 2009).

Social Acceptance and Energy Infrastructure

Over the past ten years, however, we have seen a development in which, in addition to the traditional discussions on public interests in energy supply, controversies with groups of citizens and companies are sharpening. These citizens oppose the way in which they experience the benefits and burdens of particular aspects of the current energy supply system and of the changes foreseen whether or not as part of the energy transition. This generally often involves situations where the perceived burdens affect individuals and (local) interest groups, whereas the benefits fall to society in a broader sense. Improved connections within the gas and electricity networks and with neighbouring countries facilitate the functioning of the market, with likely benefits in terms of lower prices, consumer choice, security of supply and business activity. Substantial measures are also being taken to create a low-carbon energy production on a larger scale, supplying green electricity and gas.

The consequences of these developments are the emergence of controversies around high-voltage lines, underground gas and CO_2 storage projects, on- and offshore wind farms, solar parks, shale and natural gas production, geothermal projects, and so on. This phenomenon also occurs around other infrastructures, such as those for transport, telecom, water management and water safety, and in spatial planning when it comes to the development of new residential and work locations.

In this context, the issue of *social acceptance* has come to life. It has become clear that although the construction of such infrastructural works serves a public good for all, this does not preclude resistance of concerned (groups of) local residents and of citizens in general, who reject the damage and disturbances to their living environment, nature and landscape, or question the necessity of a specific provision. Moreover, it is recognized that discussing 'social acceptance' not necessarily means that the ethically important aspects of the energy discussion are on the debating table (Taebi, 2017). In recent years, in the Netherlands, we have seen a number of delayed or failed projects, among which an underground CO_2 storage in Barendrecht and the exploration for shale gas. We also observe that, in response to these failures, operators of infrastructures to be built and governments involved are formulating their motivation in terms of usefulness and necessity. They also engage in information processes about the progress and impact of the projects, in the guidance and participation of local residents and in compensation measures. Attention to the 'management of the social environment' has become part of the public–private interaction around such infrastructures, in law and planning, in tendering conditions and in awarding permits (van de Grift et al., 2020).

This, however, does not mean that the construction of new infrastructures is now without problems. Referring to the concept of justice, it can be said that citizens frequently consider that it is unfair that they are affected in their living environment by the construction of infrastructures, although they generally support the pursuit of a sustainable and reliable energy supply at the same time. The formulation of usefulness and necessity in abstract terms, and the way in which external effects and risks are defined and allocated to specific groups of citizens, remain a source of controversy. Citizen participation and information do not always have the desired effect when it comes to reaching agreement and are often regarded as an arrangement to 'buy' acquiescence from the 'public' (Correljé et al., 2015; Cuppen et al., 2015, 2016; Taebi, 2017).

We also see the emergence of gaps between priorities at the national and the local government. Where energy objectives are formulated at national level that require the local installation of plants, factories and transport infrastructure, discussions arise between the national and local authorities about where and how these should be built. Nationally formulated values, such as the reliability and affordability of energy supply and sustainability in terms of a reduction in CO_2 emissions, clash with local values in regarding the environment, nature, safety, the related local economic interests, and ultimately with the voice of local politics (Correljé, 2017).

The Energy Transition Enters the Front Door

In respect of the energy transition, it is clear that the necessary changes to the energy system will be even more intrusive and come closer to the citizen. Of course, there are (groups of) citizens who enthusiastically embrace this perspective and take every opportunity to provide themselves with sustainable energy. However, many more citizens will face the changes in their environment, their pattern of living, their work and travel habits and their pattern of consuming with less enthusiasm. And, as stated by the Minister of Economic Affairs and Climate, Wiebes: "If the household wallet starts to suffer too much as a result of the transition, initial support will disappear quickly." (Translated from MEZK 2018).

Moreover, with the recent announcement that Dutch households will have to say goodbye to natural gas as a primary source of energy in their domestic heating and hot water supply, the energy transition enters the front door. With regard to their household wallet, their social awareness, their involvement and their comfort at home and well-being, citizens will appreciate the effects of the energy transition in rather diverse ways. And that will depend on where they live and in what kind of houses, what work they (can) do and where, what their patterns of consumption and leisure activities are, and what their (financial) capacities and possibilities for adjustment are.

Energy Justice

As stated earlier, the concept of energy justice is rooted in environmental and climate justice, where three generic forms are recognized: (1) *distributive justice*, where the question is how the benefits and burdens of energy supply and energy use are divided among groups of people; (2) *procedural justice*, which poses the question of how the decision-making process for energy supply works, who has access to and participation in it and in which way; and (3) *justice through recognition*, whereby it is stated that there may be a distinction between the way in which social, cultural, locational and other aspects can structurally influence the exposure of groups to benefits and burdens and their capacity to deal with them (Jenkins et al., 2016; McCauley et al., 2013). We therewith have an abstract framework that helps us to have an eye for who gets what (not), in what process that has been decided and whose positions have been taken into account. Over time this triple framework has been given substance and elaboration; on the one hand to increase the analytical power and to make it more explicit, and on the other to make it usable for practical policy issues.

Sovacool et al. (2016) suggest an alternative, but at the same time overlapping and complementary, framework. The three aforementioned perspectives are further elaborated on in the form of eight principles that should operationalize the concept of energy justice. These principles include, in terms of distributive justice: (1) availability, (2) affordability, (3) fairness between members of a generation (*equity*), (4) fairness between members of different generations and (5) sustainability. Regarding procedural justice, the following principles are important: (6) due process, (7) transparency and accountability, and (8) responsibility. The phenomenon of *recognition* seems to be connected with (9) due process and (10) responsibility.

Justice Assessment in the Energy System

Heffron and McCauley (2014) proposed to use the framework in the context of the entire energy system, where each segment of the energy chain is assessed from a justice perspective. The energy sub-systems generally consist of a number of vertically connected segments: production, transport, processing and consumption. Heffron and McCauley (2017: 660) underline the role of restorative justice, whereby energy justice can be created in the energy chain, in relation to the nature of the damages caused in certain segments. The idea is that attention and intervention are not only focused on punishing the perpetrator, but also on repairing damage to victims, society or nature, or proactively preventing damage. Balancing the costs of that prevention and/or recovery against the benefits could then lead to a rational termination or adjustment of the harmful activity. In fact, this way of thinking argues in terms of the economic concept of *negative external effects*, where the complete assignment of property rights to the parties involved gives rise to negotiations about monetary compensation, measures of repair, or relocation or termination of the activity (Coase, 1960).

However, the energy system and its several supply—or value—chains can be defined in many ways. The end product that is delivered to the consumer can be leading, such as the supply of oil products, gas, electricity and heat. The primary energy source can be leading, such as crude oil, coal, natural gas, nuclear energy, hydropower, wind, sun, biomass and geothermal energy, and so on. The energy service to be supplied can also be leading, such as heat, power, light, and—even—data transport. The question of how specific 'external' effects somewhere in such a chain, giving rise to occurrences of justice and injustice, can logically be assigned to the consecutive segments and the actors may cause a hop-and-skip argumentation.

Indeed, should all Dutch domestic households be forced to overhaul their energy appliances and stop using natural gas, just for the sake of reducing the earthquakes caused by the production of gas in Groningen? Should we consider to tax natural gas on the basis of CO_2 and methane emissions arising from the transport of gas imported from Siberia, or environmental damages of shale gas production in the US? Is it fair to tax domestic gas consumers while reducing electricity taxes, when most of this electricity—particularly at peak demand—is produced with gas fired power plants? Indeed, there are always many technical, economic and institutional dependencies between the segments in a system, however defined. How do we deal with the possibilities for substitution of primary energy sources, technologies and end products in the provision of essential energy services: heat, power, light and communication and data transport? Should we provide untaxed power to electric vehicles?

And even with regard to services, there is a degree of substitution potential, for example when it comes to (tele) communication versus transport or the use of energy to construct low-energy houses versus heating those houses with gas. Either of those alternatives will use energy of whatever origin, with particular consequences and effects. As a consequence, it makes little sense to simply link particular situations of injustice of any kind in up- or downstream segments of a supply chain, to actions and interventions in either the use and consumption sphere, or the production segment. The problems are almost always much more complex.

Energy Justice in a Multi-level Framework

An alternative approach in applying the notion of energy justice is suggested by Jenkins et al. (2018), with the multi-level perspective approach (MLP) of the socio-technical system serving as a framework (Cherp et al., 2018; Geels, 2002). Bouzarovski and Simcock (2017) and Sovacool et al. (2019) take a similar approach to which they add the notion of space; identifying injustices at the scale of the community, the nation or region, or the global scale. In the MLP context, occurrences of energy (in)justice are linked to three different levels: the *niche*, the *socio-technical regime* and the *landscape*. It is the interaction between developments at the level of the niches and the socio-technical system (and within that), and in the context of the landscape, that transitions are given shape.

At *the level of the niche(s)* we find concrete, more or less innovative, applications of technologies or systems under development, such as electric cars, individual or neighborhood batteries, or biogas installations. Developments in niches are dynamic and their embedding in technical and institutional frameworks has often not crystallized yet. A justice perspective applied at this level should make it possible to identify potential sources and forms of injustice at an early stage. Technological adjustments can be proposed and assessed with these insights. Aspects of an appropriate institutional embedding, in terms of rules of conduct, norms and standards, can be explored, with which social acceptance can be strengthened. We see here a possible application of concepts such as socially responsible innovation (MVI) (Taebi et al., 2014). Nevertheless, we stress that at the niche level it is impossible to make a full evaluation of the institutional embedding of such new technologies. This only comes to light at the level of the socio-technical regime and larger scale implementation, when issues of economic, market, technical, social and system coordination become important and have to be addressed.

At *the level of the socio-technical regime*, the established technological systems, their institutional embedding, the resulting routines and practices and their social effects are examined. The regime creates stability and gives direction to further technological developments and to the behavior of public and private actors. Changes in the regime take place under the influence of the dynamics within the regime and as a result of developments in the niches, also influenced by landscape shifts. At the regime level, as argued by Jenkins et al. (2018), energy justice can play a role in mapping and evaluating the social, economic and ecological effects of the functioning of (parts of) technical systems, such as the electricity or gas supply infrastructures, district heating networks, wind parks, electric vehicle loading systems, etc. The establishment of normative criteria and assessment frameworks can help policy makers and companies to assess the functioning of those systems, as well as the possible changes therein. Here it can be checked to what extent such systems meet the social requirements in terms of distributive and procedural justice, and of justice through recognition regarding the impact on those involved.

The third level of the MLP concerns *the macro landscape* (Jenkins et al., 2018: 70). Here we find the embedding of actors and institutions in a relatively stable social and global context of political, social and cultural values, including knowledge and scientific insights. The landscape level in the MLP literature is usually considered static and inhibitory or facilitating. However, here we also see elements that sometimes change relatively quickly and thus influence the notion of energy justice and its application. Examples are the way in which the behavior of multinational companies and the role of the state in the economy is evaluated. It also may concern international relations, developments in the oil and gas market and, for example, the consequences of the nuclear disaster in Fukushima. Other shift parameters include the development of new knowledge and insights into the effects of energy use on climate change and the consequences thereof. Such phenomena influence the identification and societal and political assessment of aspects of energy justice. These, in turn, influence how the argumentation and evaluation is conducted at the lower two levels, giving rise to shifts within the regimes and to innovation in (new) niches.

It can be argued that the positioning of energy justice in relation to the goals of affordability, reliability and sustainability partly takes shape at the level of the landscape. Examples are the expectation of higher oil prices in the future due to depletion and the power of OPEC, the risks of EU gas dependence on Russia, the hazards of nuclear energy, the expected consequences of global warming and the deterioration of the Arctic by oil and gas extraction, and so on.

Identification of Claims of Energy Justice

With regard to each of the three levels, the question can of course be asked how claims of energy (in) justice can be identified, and whether or not they will have an effect in concrete policies or strategies (Pesch et al., 2017a). Building on the above, Pesch et al. suggest an approach in which the role of controversy and conflict around energy (projects) is central. Controversy is seen as an indicator to identify injustices and helps to understand how such claims, either or not, are articulated and accepted as a relevant public value. To this end, a distinction is made between, on the one hand, the legally established formal evaluation process, be it in the form of macro-economic or environmental models, or as (Societal) Cost Benefit Analysis, Environmental Impact Assessment and licensing and planning procedures. On the other hand, however, there is an informal social process, which takes place in the public discourse and can take many forms. The emergence and growth of public protest about particular forms of energy (projects) can be seen as an (alleged) lack of attention to certain social values in the formal process (Pesch et al., 2017b).

In the public and political debate, such values may be articulated and then be included in the formal policy process, or not. In the occurrence of such controversies there are three characteristic differences between the two trajectories. Firstly, in the way in which values are expressed, the formal process usually involves a legally and technically defined rationality, whereas stories and shared experience and feelings are central to the informal process. This results in diverging claims for justice. Secondly, procedural justice is dominant in the formal process, determining the way in which recognition and distributive justice are taken into consideration. On the other hand, the justice of recognition is often central to the informal process, distinguishing how different societal groups are affected to their own feeling. Thirdly, the moral conviction in both processes is based on different democratic principles. On the one hand there is the formality of institutionally guaranteed and legally established rules from which parties derive rights, and on the other hand it is often about moral self-determination of citizens who belong to a specific community (Pesch et al., 2017a).

By not considering both processes as separate, but by seeing them in relation to each other, (possible) injustices can be identified, understood and discussed. That may, or may not, lead to adjustments. This may enable us to use (in)justice as a useful concept with which controversies around the energy supply can be understood and assessed. Energy justice thus may become, as Sovacool and Dworkin (2014: 20) state: "an appropriate orientation for considering, balancing and prioritizing various justice claims that arise in energy patterns and decisions". This, however, is not a straightforward exercise, as will be shown in the following section.

Values in the Energy Transition

The Landscape Level

In the Netherlands, citizens are increasingly confronted with the effects of the energy transition, as agreed in the Paris agreements and the European objectives in the area of CO_2 emission reduction, and elaborated in Dutch policy measures. Here the global objective is translated into a European timetable of national emissions, the Green Deal. It is then implemented by the EU member states as more or less concrete, sectorally oriented, transition targets at the national level. Knowledge development and changing insights into CO_2 emissions and climate change at the level of the landscape give rise to justice claims with regard to the existing energy regime(s). It is clear that a complex set of values, related to the effects of global warming, is linked to the nature and structure of the Netherlands' energy system.

The Socio Technical System Level

Increasingly, policy measures are being taken to make energy supply more sustainable in terms of CO_2 emissions, such as in the Dutch Energy Agreements and the Regional Energy Strategies. Examples at the system level are the construction of wind farms and solar fields, the closure of the older coal-fired power stations and the reduction of the role of natural gas in the energy mix. In recent years there has also been an increase at the local, municipal, level of initiatives for sustainability in the built environment, the transport sector and in the energy consumption of the public sector (Weijnen et al., 2015). But when implementing this policy, in particular in the form of wind and solar parks, we also see that the loss of all kinds of local values is questioned as being *unjust*.

In parallel, a second important shift is taking place at the system level. Since the mid-1960s, based on the discovery of a huge gas field in Groningen, natural gas has evolved as the main source of energy fuelling Dutch households and economic production (Correljé et al., 2003). As from the early 1990s, the province of Groningen was hit by earthquakes as a result of the production of gas from the field that extends under a large part of the province. Over time, with the pressure in the depleting field decreasing, those earthquakes have augmented in number and force. This gave rise to increasingly powerful protests from the inhabitants of Groningen, who found themselves supported by their local politicians and later also by national politics. The deterioration of safety and the destruction of property led to justice claims that related not only to the distribution of benefits and burdens of gas production, but also— possibly even more—to the long-awaited recognition (and even the initial denial) of the relationship between gas production and earthquakes, and the consequences for the inhabitants. Lack of procedural justice is also generating fierce criticism. This involves both the decision-making process by the Minister of Economic Affairs concerning the scale of the annual production of gas, as well as execution of the compensatory procedure for damages, as legally provided. Recently, the notion of flawed procedural justice has also been applied in respect of the procedures and implementation of a programme by which existing houses and buildings will be reinforced, to withstand possible future earthquakes (van den Beukel & van Geuns, 2019; Van der Voort & Vanclay, 2015).

These justice claims and their political articulation led to action. Gas extraction was thus reduced in a few steps. The formal motivation for this lies in a number of "recommendations" from the State Supervision of Mines (SodM) and various investigations into the trade-off between gas production, earthquake risk and the security of gas supply. Nevertheless, the Council of State ruled in November 2017 that the minister had to take a new and better substantiated decision. The *risk* to the people in the earthquake zone was not sufficiently taken into consideration in the justification. Nor was it sufficiently motivated why *security of supply* was taken as the lower limit for the amount of gas to be extracted, despite the uncertainty about the consequences. Moreover, it was not made clear what measures are actually feasible to limit the need for a specific volume gas (Raad van State, 2017).

After the unexpectedly severe earthquake at Zeerijp in January 2018, the SodM recommended a production of 12 billion m^3 per year. In March, however, the government announced that the gas production in the Groningen field would have to be terminated as quickly as possible, to avoid a further increase of the earthquake risk, and to restore the perception of safety for the inhabitants. This implies that the field will not be fully depleted. By the end of 2022, gas extraction must have fallen to below 12 billion m^3 per year. Depending on the effect of the measures to reduce gas consumption, a decrease is expected from October 2022 to 7.5 billion m^3 and possibly less. Moreover, by 2022 all 170 large-scale industrial consumers of Groningen gas must have switched to high-calorific gas or alternative energy sources. Also, with foreign buyers of Groningen gas, in northern Germany, France and Belgium, arrangements are struck to accelerate the reduction of their gas consumption. After 2022, gas extraction will be further reduced to zero (MEZK, 2018; Beukel & Geuns, 2020; Beukel & Beckman, 2019).

At the level of the socio-technical system, at first sight, we see a policy that is inspired by securing the value of *solidarity* and *safety* for the people of Groningen. Indeed, the right to extract gas from the Groningen field by the NAM, a joint venture of Shell and Exxon-Mobil, as agreed with the Dutch State and laid down in a policy paper in 1962, is severely restricted. In the second instance, however, we see a different, much more complex, pattern of values at stake. Firstly, alongside a reduction in the

Groningen production, it is quite possible to import high-calorific gas from Norway, Russia or elsewhere in the form of LNG. However, this requires an investment by Gasunie in additional transport infrastructure and the construction of a nitrogen plant to "dilute" that gas to the quality of Groningen gas. The costs of this are borne by all gas consumers in the Netherlands, given the current mechanism of cost socialization. This implies, at the regime level, a weakening of the value of *affordability* via the regulated transport tariffs, in particular for those consumers who must continue to use gas because they have no alternative. Secondly, it means that more gas has to be imported into the Netherlands and therefore into Europe, which can be seen as an impairment of the value of *energy independence* or *reliability*, especially with regard to Russia. In the current international political context in Europe, this is a difficult issue at the landscape level. Thirdly, it is often said that the pace at which the Netherlands could switch off from gas—as a fossil fuel—would be delayed, when foreign gas would be imported to replace Groningen gas. This would put the value of *sustainability* at stake in the longer term, depending on how that Groningen gas is going to be replaced by alternatives. And, on the shorter term, sustainability will be jeopardized by larger CO_2 and methane emissions associated with gas imported from Russia, or as liquefied shale gas from the US.

The Niche Level

The initial plans to reduce the gas production in Groningen, in combination with the pursuit of CO_2 emission reduction, had already convinced a number of Dutch municipalities that they should voluntarily say goodbye to natural gas. What certainly contributed to this was the growing criticism of natural gas as a source of energy and of the governance of the gas industry in the Netherlands, also inspired by the protests against shale gas development (Correljé, 2017). Nevertheless, the decision to phase out gas production in Groningen before the field would be depleted requires a significant acceleration of this conversion. Newly constructed and renovated buildings will have to be (re)constructed gasless. Over the slightly longer term, a gradual disconnection of gas will have to take place in the existing built environment. To that end, municipalities are committed to developing regional transition plans.

Therewith, as stated above, the energy transition crosses the threshold of the frontdoor. From a technical point of view, citizens are now confronted with an uncertain action perspective regarding alternative heating solutions. Technologically speaking, heat pumps are still in their infancy compared to the current high-efficiency gas boilers. Heat networks still have a long way to go in terms of their institutional embedding, possible business models and their technical development. In addition, it is clear that a large variation in the living environment and types of housing will lead to major differences in suitability and switching costs with regard to new forms of heating and thermal insulation. What also seems important here is that from a relatively homogeneous situation, in which energy similar to water and the sewage services is provided at standardized conditions, we may see a rapid shift to a much larger diversity in supply conditions, depending on the specific circumstances of the individual users.

This is also the case when it comes to IT facilities. What makes a difference here is that the telecom sector (fixed telephony, cable companies and mobile telephony) has now gone through a relatively long period of competitive technology and market development. So a variety of more or less similar solutions is available. Yet, there are still significant (local) differences in the quality and costs of provision.

Generally, the role of smart grids, ICT and digital platforms is seen as a great promise to facilitate a new, smart, sustainable and efficient provision of energy services. That could well be, given the opportunities of ICT-based platforms in coordinating supply and demand, the allocation of production, transport and storage capacities, and the allocation of costs and benefits to the users of those smart energy systems. Nevertheless, there is an important point of attention as such systems can process the collected information in all sorts of ways. The information and coordinative mechanisms provided can be used not only very smartly, but also shrewdly and strategically, to profitably discriminate among users. Depending on the governance of such smart grids, discrimination among the various groups of users can take place, depending on their capabilities to act and react (Van Dijck et al., 2016). It is obvious that justice issues of recognition and distribution are at stake here that require attention.

It is obvious that the course of the transition at the local level, in terms of costs, quality and comfort for citizens and businesses, and regarding the process itself, will be highly dependent on the ownership relations, the capacity and the cooperation of municipalities, network managers, project developers, housing associations, the construction sector and installation companies, and any new parties. It is already clear that this will lead to highly varying circumstances in different municipalities, where the size and capacity and local politics will be determining factors in their effectiveness in coordinating a 'just' transition.

In addition to the direct consequences of the transition for the energy supply of citizens and companies, it is to be expected that radical second-order effects will occur. Residents of local communities, neighborhoods and municipalities will be confronted with the construction of new energy systems, new technologies and new infrastructures, with new effects on their living environment. Citizens will have to adjust their travel behavior when it comes to commuting, necessary journeys for education, medical and other services, and leisure activities. As employees in particular economic sectors, they may be confronted with radical changes in business processes, or possibly even with the termination thereof. This will be accompanied by new activities, that place new demands on training and knowledge development. On a local scale this can have important positive or negative influences on (the structure of) employment and the supply of labor (see Kooger et al., 2017; SER, 2018).

Citizens will tend to judge the consequences of these changes in terms of justice. Above it has been stated that large differences may arise between groups of people, depending on where they live and in what type of houses, their work, their patterns of consumption and leisure activities, and their (financial) capacities and possibilities for adjustment. This involves *distributive justice* in respect of the question of how the benefits and burdens of new forms of energy supply and use are distributed among those groups of people. *Justice through recognition* seems necessary to gain access to those different groups of citizens and to get a picture of what consequences they will experience from the changes. From there, it can be considered which specific approach is most suitable, with regard to the technologies to be used, the means for financing, providing support and information, and so on. *Procedural justice* also requires understandable, foreseeable, decision-making and implementation processes that facilitate insight, access and participation where necessary in a credible and consistent manner.

Conclusion

How can the concept of justice contribute to a socially responsible policy for the energy transition? Sovacool and Dworkin (2015) argue that it allows us to link different issues of justice around energy. As an analytical approach, it could help researchers to identify the various relevant values at stake in the system of energy provision. This can help policymakers to make informed choices. To do so, it appears to be of great importance to consider the nature of the socio-technical system of energy provision, its functioning, and its specific local layout.

Above we have demonstrated that the evolution of the energy system is increasingly driven by sustainability goals at the level of the landscape in the form of CO_2 emission reduction, and specifically for the Netherlands' energy system, the decision to reduce the gas production in Groningen. This will lead to radical changes in the energy supply, which will have both direct and indirect consequences for citizens and businesses. The advancing energy transition is likely to show a wide variety in the consequences experienced by citizens and businesses, depending on their specific circumstances. We expect to see great contrasts in their ability to respond to these changing conditions in energy use and provision.

The concept of energy justice provides a starting point in terms of the distinction between distributive and procedural justice and justice through recognition. In particular, the recognition of the major differences in benefits and burdens and in the perspective for action between groups of citizens appears as an essential element for a socially responsible transition process and for the selection of suitable policy instruments. The suggestion, however, that justice issues can be identified and solved at the various niche, regime and global levels, is too simple. Solving such issues will always cause new value conflicts and situations of injustice at and between the different levels.

There is no straightforward way to avoid conflicts. However, understanding these conflicts can be helpful in identifying injustices and concretising values that appear to be compromised. At the niche level, a justice perspective applied to new options for energy supply in a particular environment makes it possible to identify possible sources and forms of injustice at an early stage. Technological adjustments and social aspects can be proposed and discussed with these insights.

At the system level, aspects of an appropriate formal institutional embedding can be explored, in terms of the rules of the game, financial and economic coordination, norms and standards, and planning and phasing. This is a learning trajectory in which the experiences of actors, the effects of upscaling and the associated institutional and technological development may gradually lead to new insights and possibilities.

A particular challenge lies in dealing with the characteristic differences between the way in which values are expressed and used as a justice claim in formal and informal valuation processes. It is clear that justice of recognition must be a crucial aspect of the interaction. In addition, some sensitivity in understanding and interpretation will be needed to translate the stories, experiences and feelings of citizens, but also those of public and private organizations, into the values that must be taken into account in decision-making and in the institutional embedding of the energy transition (see also Jenkins et al., 2020). There is no doubt about the need for procedural justice. That has been clearly demonstrated in the Groningen case, where the faltering approach to recognition, recovery and compensation has contributed to institutionalized mutual mistrust between residents, local and national government and the gas industry. A perceived lack of procedural justice seems a guarantee that the moral self-determination of citizens in their community will turn against the energy transition and the authorities and businesses involved. In a context in which all institutions, knowledge, considerations and technologies will be questioned anyway, it then becomes difficult to reach any kind of workable consensus.

References

Bouzarovski, S., & Simcock, N. (2017). Spatializing energy justice. *Energy Policy, 107*, 640–648

Cherp, A., Vinichenko, V., Jewell, J., Brutschin, E., & Sovacool, B. K. (2018). Integrating techno-economic, socio-technical and political perspectives on national energy transitions: A meta-theoretical framework. *Energy Research & Social Science, 37*, 175–190

Coase, R. (1960). The problem of social cost. *Journal of Law and Economics, 3*, 1–44

Correljé, A. (2017) The Netherlands: Resource management and civil society in the natural gas sector. In Indra Overland (Ed) *Public brainpower: Civil society and natural resource management.* Pallgrave/Macmillan, London.

Correljé, A., Van der Linde, C., & Westerwoudt, T. (2003). *Natural gas in the Netherlands. From cooperation to competition?*

Correljé, A., & Verbong, G. (2004). The Transition from coal to gas: Radical change of the Dutch gas system. In B. Elzen, F. Geels, & K. Green (Eds.), *Break on through to the other side: Technological transitions to sustainability through system innovation* (pp. 114–134). Cheltenham UK, Edward Elgar.

Correljé, A. F., Cuppen, E., Dignum, M., Pesch, U., & Taebi, B. (2015). Responsible innovation in energy projects: Values in the design of technologies, institutions and stakeholder interactions. *Responsible Innovation, 2*, 183–200

Cuppen, E., Brunsting, S., Pesch, U., & Feenstra, Y. (2015). How stakeholder interactions can reduce space for moral considerations in decision making: A contested CCS project in the Netherlands. *Environment and Planning a, 47*, 1963–1978

Cuppen, E., Pesch, U., Taanman, M., Remmerswaal, S. (2016) Normative diversity, conflict and transitions: shale gas in the Netherlands. *Technological Forecasting and Social Change, 29.*

de Jong, J. J., Weeda, O., Westerwoudt, T., & Correljé, A. F. (2005). *Dertig jaar Nederlands energiebeleid: Van bonzen, polders en markten naar Brussel zonder koolstof*. Clingendael International Energy Programme.

Geels, F. W. (2002). Technological transitions as evolutionary reconfiguration processes: A multi-level perspective and a case-study. *Research Policy, 31*, 1257–1274

Goudsblom, J. (2001). *Vuur en beschaving*. Ooievaar.

Groenewegen, J., & Correljé, A. (2009). Public values in utility sectors; economic perspectives. *International Journal of Public Policy, 4*(5), 395–413

Heffron, R. J., & McCauley, D. (2014). Achieving sustainable supply chains through energy justice. *Applied Energy, 123*, 435–437

Heffron, R. J., & McCauley, D. (2017). The concept of energy justice across the disciplines. *Energy Policy, 105*, 658–667

Hesselmans, A. N., & Verbong, G. P. J. (2000). Schaalvergroting en kleinschaligheid: de elektriciteitsvoorziening tot 1914', blz. 124–139 in J. Schot, H. Lintsen & A. Rip *Techniek in Nederland in de twintigste eeuw. Deel 2. Delfstoffen, energie, chemie*, Eindhoven: Walburg Pers.

Hesselmans, A. N., Verbong, G. P. J. & Buiter, H. (2000a). Binnen provinciale grenzen: de elektriciteitsvoorziening tot 1940, blz. 140–159 in J. Schot, H. Lintsen & A. Rip *Techniek in Nederland in de twintigste eeuw. Deel 2. Delfstoffen, energie, chemie*, Eindhoven: Walburg Pers.

Hesselmans, A. N., Verbong, G. P. J. & van den Berg, P. (2000b). Elektriciteitsvoorziening, overheid en industrie 1949–1970', blz. 220–237 in J. Schot, H. Lintsen & A. Rip *Techniek in Nederland in de twintigste eeuw. Deel 2. Delfstoffen, energie, chemie*, Eindhoven: Walburg Pers.

Jenkins, K. (2018). 'Setting energy justice apart from the crowd: Lessons from environmental and climate justice'. *Energy Research & Social Science, 39*, 117–121

Jenkins, K., McCauley, D., Heffron, R., Stephan, H., & Rehner, R. (2016). Energy justice: A conceptual review. *Energy Research & Social Science, 11*, 174–182

Jenkins, K., Sovacool, B. K., & McCauley, D. (2018). Humanizing sociotechnical transitions through energy justice: An ethical framework for global transformative change. *Energy Policy, 117*, 66–74

Jenkins, K. E., Stephens, J. C., Reames, T. G., & Hernández, D. (2020). Towards impactful energy justice research: Transforming the power of academic engagement. *Energy Research & Social Science, 67*, 101510

Kooger, R., Straver, K., & Rietkerk, M. D. A. (2017). *Essay bundel 'De ethiek van de energietransitie': Inleidende essays over de winnaars en verliezers van de energietransitie*. Energieonderzoek Centrum Nederland.

McCauley, D., Heffron, R., Stephan, H., & Jenkins, K. (2013). Advancing energy justice: The triumvirate of tenets. *International Energy Law Review, 32*(3), 107–110

Millward, R. (2005). Private and public enterprise in europe energy, telecommunications and transport, 1830–1990 Series: Cambridge Studies in Economic History—Second Series, Cambridge.

MEZK (2018) Kst. 32 813 Nr. 163 *Brief van de Minister van Economische Zaken en Klimaat aan de Voorzitter van de Tweede Kamer der Staten-Generaal*, Den Haag, 23 februari 2018.

Pesch, U., Correljé, A., Eefje, C., & Taebi, B. (2017a) Energy justice and controversies: Formal and informal assessment in energy projects. *Energy Policy, 109*, 825–834.

Pesch, U., Correljé, A., Cuppen, E., Taebi, B., & van de Grift, E. (2017b). Formal and informal assessment of energy technologies. In J. van den Hoven, E. J. Koops, T. Swierstra, H. Romijn, & L. Asveld (Eds.), *Responsible Innovation* (Vol. 3). Springer.

Raad van State (2017) Uitspraak 201608211/1/A1. Raad van State, s'Gravenhage. https://www.raadvanstate.nl/@109356/201608211-1-a1/. Accessed on January 07, 2020.

Rooijendijk, C. (2009). *Waterwolven: Een geschiedenis van stormvloeden, dijkenbouwers en droogmakers*. Atlas.

SER. (2017). *Governance van het energie- en klimaatbeleid*, SER Advies nr. 5, april 2017, Den Haag: Sociaal-Economische Raad.

SER. (2018). *Ontwerpadvies Energietransitie en werkgelegenheid*, Bestemd voor de raadsvergadering d.d. 19 April 2018, Den Haag: Sociaal-Economische Raad.

Sovacool, B. K., & Dworkin, M. H. (2014). *Global energy Justice: Problems, principles, and practices*. Cambridge University Press.

Sovacool, B. K., & Dworkin, M. H. (2015). Energy justice: Conceptual insights and practical applications. *Applied Energy, 142*, 435–444

Sovacool, B., Heffron, R. J., McCauley, D., & Goldthau, A. (2016). Energy decisions reframed as justice and ethical concerns. Nat Energy 1https://doi.org/10.1038/nenergy.2016.24

Sovacool, B. K., Hook, A., Martiskainen, M., & Baker, L. (2019). The whole systems energy injustice of four European low-carbon transitions. *Global Environmental Change, 58*, 101958

Taebi, B. (2017). Bridging the gap between social acceptance and ethical acceptability. *Risk Analysis, 37*(10), 1817–1827

Taebi, B., Correljé, A., Cuppen, E., Dignum, M., & Pesch, U. (2014). Responsible innovation as an endorsement of public values: The need for interdisciplinary research. *Journal of Responsible Innovation, 1*(1), 118–124

van de Grift, E., Cuppen, E., & Spruit, S. (2020). Co-creation, control or compliance? How Dutch community engagement professionals view their work. *Energy Research & Social Science, 60*, 101323

van den Beukel, J., & Beckman, K. (2019). The great Dutch gas transition. Oxford Institute for Energy Studies, Oxford Energy Insight: 54, July 2020.

van den Beukel, J., & van Geuns, L. (2020). Groningen gas: the loss of a social license to operate. The Hague Centre for Strategic Studies, HCSS geo-economics, januari 2020.

van der Linde, I. (2008). *De slag om de Waddenzee: Een terugblik op vijf jaar politieke strijd*. IMSA.

Van der Voort, N., & Vanclay, F. (2015). Social impacts of earthquakes caused by gas extraction in the Province of Groningen, The Netherlands. *Environmental Impact Assessment Review, 50*, 1–15

van Dijck, J., Poell, T., & de Waal, M. (2016). *De platformsamenleving: Strijd om publieke waarden in een online wereld*. University Press.

Weijnen, M., Correljé, A. & de Vries, L. (2015). *Infrastructuren als wegbereiders van duurzaamheid*, Working Paper nummer 12, Den Haag: Wetenschappelijke Raad voor het Regeringsbeleid (WRR).

The Hidden Dimension of the Energy Transition: Religion, Morality and Inclusion—A Plea for the (Secular) Sacred

Maarten J. Verkerk and Jan Hoogland

Abstract This chapter explores the energy transition from a philosophical perspective. We argue that there is a hidden dimension in the current discussions about sustainability. This hidden dimension can be found first of all in the fact that phenomena such as the denial of global warming, the rise of populism and the increase in social contradictions are not seen in their context. At a fundamental level, it appears that all these phenomena are characterized by broken connections: man no longer feels connected with the Transcendent, the human being and the planet. On the basis of the above analysis, we outline some action perspectives. We conclude that the energy transition not only requires addressing technological, economic, social and legal problems, but that moral and religious aspects must also be discussed. Because it is precisely religious or moral values that motivate and inspire people to strive for an inclusive energy transition and release a lot of creative energy.

Life-Size Dilemmas

The issue of climate change brings enormous dilemmas. On the one hand, the reports of the IPCC provide compelling evidence that climate change is caused by human acts and that drastic measures are required to limit global warming (IPCC, 2013). On the other hand, support for climate policy is dwindling among large sections of the population, populist parties that deny the existence of climate change and/or deny that this change is caused by human acts are on the rise, and finally there is the cancellation of the Paris (2015) UNFCCC climate agreements by the United States of America.[1] The sustainability issue is so great that unanimous support is

---

[1] Restored on the first day of the Biden administration.

M. J. Verkerk (✉)
Faculty of Arts and Social Sciences, Technology and Society Studies,
Maastricht University Science, Maastricht, The Netherlands
e-mail: maarten.verkerk@home.nl

J. Hoogland
University of Twente, Enschede, The Netherlands

M. P. C. Weijnen et al. (eds.), *Shaping an Inclusive Energy Transition*,
https://doi.org/10.1007/978-3-030-74586-8_4

needed to achieve feasible solutions in the relatively short term. However, there is no consensus, either nationally or globally. The division in society is so deep that sustainability seems to become an insurmountable problem.

The philosopher Bruno Latour suggests in his book Down to Earth (2018) that the overwhelming dilemmas and the insolvability of the climate issue stem from a 'hidden dimension' in our thinking and our behaviour. He positions that hidden dimension in our scientific attitude: we have learned to place the earth at a distance and to look at it from a distance. As a consequence, we do not feel connected with the earth. This leads to indifference to the alarm systems that warn us for the climate crisis and the sirens that have been blaring full fast about global warming. Latour states that the enormous dilemmas related to our climate can only be overcome by addressing the hidden dimension in our culture: our connection with the earthly. Furthermore, he suggests that this hidden dimension has a religious character (Latour, 2017, p. 193 ff.).

Latour asserts that we can understand nothing about the politics of the last 50 years if we do not put the question of climate change and its denial 'front and centre' (Latour, 2018, p. 2). He argues that we have entered into a 'New Climatic Regime'. That means, we have arrived in a situation in which the earth changes under influence of the activities of mankind. In other words, our planet is not anymore a passive background but has become an actor that plays its role on the world stage. Latour emphasizes that the emergence of a New Climatic Regime is also evident from the increase in all kinds of social phenomena. In his view, the explosion of social inequalities and the rise of populism are symptoms from that new regime (Latour, 2018, p. 2). He argues that the ruling classes—'the elites'—have decided that it is pointless to act as though history were going to continue to move toward a 'common horizon', toward a world in which 'all humans could prosper equally' (Latour, 2018, p. 1).

If Latour's analysis is correct, then the climate crisis and its denial are related to (1) religious choices and (2) social inequalities. This implies that the concept and goal of an inclusive energy transition could be a very controversial one. After all, the present dialogues about the energy transition are about technology, political decisions, policy decisions, management of change, and so forth. The idea that the energy transition is also about religious choices, morality, and fighting social inequalities is highly provocative. In this chapter, we explore these provocative ideas.

This chapter has the following set-up. In Section "Latour: The Hidden Dimension" we explore the trail of the 'hidden dimension' as proposed by Bruno Latour. We argue that the hidden dimension in one way or another is related to philosophical and religious questions about the relationship between man, fellow man, and nature. In Sections "The Idea of Broken Connections" and "The Question of the Sacred" we delve into the hidden dimension on the basis of the work of philosophers Luc Ferry and Bronislaw Szerszinsky. They make it plausible that this dimension has an existential nature and can be related to the ideas of religion, worldview and the 'sacred'. In Section "The Hidden Dimension in the Worlds of Engineers, Policy Makers, and Politicians" we argue that the 'hidden dimension' comes to the fore in the values of engineers, policy makers, and politicians, in the interests of stakeholders, and in the ideals and basic beliefs in society. In Section "A Plea for the (Secular) Sacred" we discuss the fundamental questions about the existence of human beings in this world. It is about human connections, about what transcends people, and about renewal of human being and society. We advocate for a value system that transcends human being and/or for a (secular) sacred that unites humanity. We close with some remarks about action perspectives.

Latour: The Hidden Dimension

Latour (2017, 2018) wonders why climate scepticism can be rampant despite an abundance of scientific evidence. He also wonders how a ruler like Donald Trump can step out of the international climate agreement. According to Latour, Trump makes us aware that a real fight is going on in which the climate issue plays a central role. According to Latour (2018, p. 5), there are two options. The first is to deny that something is going on and fight for survival of the members of your own clan, if necessary, at the expense of the rest. The second is to change course radically and to revise fundamentally the relationship of human being to the planet. For Latour, denial is not an option, so he chooses to face the challenge. That option, however, imposes challenging demands on those involved and in particular on those who are scientifically engaged with the climate issue and who have a great deal of insight into it.

Since enlightenment, we have always assumed that humanity could succeed in increasing its knowledge of the world and could use it to steer development towards prosperity, happiness and well-being. In an increasingly rational organization of the world, local interests would increasingly give way to a global order in which the public interest would be leading. Latour argues that the climate problem clearly shows that this line of thinking is too simple. After all, the natural world can no longer handle humanity's growing claims and 'strikes back'. He expressively describes the change in the relationship between man and nature by using the metaphor of a stage and its actors. He writes: "Humans have always modified their environment, of course, but the term designated only their surroundings, that which, precisely, encircled them. They remained the central figures, only modifying the decor of their dramas around the edges. Today, the decor, the wings, the background, the whole building have come on stage and are competing with the actors for the principal role. This changes all the scripts, suggests other endings. Humans are no longer the only actors, even though they still see themselves entrusted with a role that is much too important for them" (Latour, 2018, p. 43). In other words, nature is no longer the objectively recognizable, technically accessible and economically available environment in which people shape their own life and culture, but has itself become a player. This is evidenced by the changes in our environment: global warming, the melting of glaciers and ice caps, rising sea levels, an increasing frequency of extreme weather events with devastating consequences, and the depletion of natural resources.

How did it get so far that nature strikes back? That nature has become a political actor? Latour (2018, pp. 64–72) blames modern science for paving the path to the New Climatic Regime. Science has developed a 'Global' approach: in the way it pursued the quest for knowledge it put nature at a distance and neglected the interactions between mankind and its natural environment. Scientists are trained as objective and rational observers who are 'external to the social world' and 'indifferent to human concerns'. Precisely this external and indifferent approach nourishes the climate scepticism of action groups like the 'yellow jackets': people with little influence on the climate problem, who are nevertheless the victims of the explosion of inequalities and have to bear the burden of the climate problem. According to Latour, we will have to look at nature in a different way. Namely, no longer as a passive body that allows rational manipulations by people, but as part of a complex whole that plays an active role in the creation of sustainability—which depends on a range of critical balances interacting through various mechanisms in the whole of a complex adaptive system. In his book Facing Gaia (2017), Latour calls this complex whole 'Gaia', and in his more recent book Down to Earth (2018) he speaks of the 'terrestrial'. Latour believes that the way scientists deal with the planet requires a new mindset. Especially, the mentality that nature is 'sensitive to human actions' (Latour, 2018, p. 67).

More generally, Latour (2017, pp. 206–208) reproaches modern man for his belief that the Apocalypse has already taken place and that the 'Promised land of Modernity' has already been reached. He contends that modern man hears the alarm systems about the climate crisis. He argues that modern man, deep down, does not acknowledge the climate crisis and does not believe that a change in his way of life is inevitable. Therefore, Latour believes that the origin of climate scepticism lies not in a lack of solidity of our knowledge and understanding of nature but stems from our own existential position in nature. We cannot accept that the 'Promised Land of Modernity' has not arrived. We cannot accept that nature strikes back and that the whole scene has changed.

If we let the previous thoughts sink in, the contours of the 'hidden dimension' of the climate crisis become increasingly clear. The *first* contour is found in the interpretation of the relationship of man with his fellow human beings and the relationship of man with nature. These relationships are characterized by terms such as 'external', 'indifference', and 'detachment'. The *second* contour is found in the existential interpretation of humanity. In this interpretation, terms like 'religious origin', 'Promised Land', 'Apocalypse', and 'Gaia' come to the fore.

The Idea of Broken Connections

How to understand the changing relationship of man with fellow man and man with nature? To answer this question, we have to dig deeper in the history of western philosophy. Luc Ferry's book *Learning to Live. A User's Manual* (2010) is used as a guide here. Ferry believes that we need philosophy to 'understand the world we live in' and to 'live a better and freer life'. He shows that in the course of history different

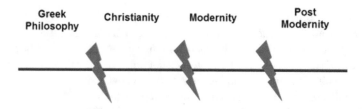

Fig. 1 Different connections of man with fellow men and nature

philosophies or 'manuals' have been provided to learn to live, as will be discussed below. Ferry's objective is to make the starting points of these manuals explicit and to describe the challenges for today.

Ferry tries to understand the history of our culture by asking three questions:

(1)   How to understand reality?
(2)   How to live?
(3)   How to find salvation?

The first question is related to the (perceived) order in our reality, the second question to morality, and the last question to wisdom and the meaning of life. The questions of Ferry are related to the main questions of the well-known philosopher Immanuel Kant: 'What can I know?' 'What should I do?' and 'What can I hope?'

Ferry runs through the history of philosophy with big steps. He distinguishes four main traditions: Greek philosophy, Christian philosophy, modern philosophy, and post-modern philosophy, see Fig. 1. He tries to understand every tradition from within by using the three previously asked questions as signposts. Ferry asks himself whether the history of philosophy can be interpreted as a history of continuity or has to be understood as a history of discontinuity. Let us first follow the road marked by Ferry.

Ferry discusses the Greek philosophy on the basis of the thinking of the Stoics. The Stoics describe the cosmos as a living organism or a giant creature. Every organ fulfils a beautiful function and cooperates harmoniously with the other organs. The Stoics believe that the order of reality is a 'divine order.' The order of the universe is also rational, consonant with what the Greeks called 'Logos'. The Stoics furthermore believe that the cosmos is not only harmonious but also just and good. As a consequence, the answer to the question 'How to live?' is that every person has to live in agreement with the divine order of society. Every human being has to take his or her own position in society. That means, the hierarchical relationships of masters and slaves, males and females, and Greeks and barbarians have to be respected. In Stoic thinking, death does not mean a definitive end, but is a transition to another state. Here we find an answer to the last question about salvation: man will be united with the cosmic order.

Ferry shows that the rise of Christianity implies a radical break with Greek thinking. Firstly, this break is evident in the different understanding of the order of reality. The Stoics believed that the Logos, the divine principle, was identical to the harmonious order of the world. Christians, however, identified the Logos with a unique person: Jesus Christ, the Logos incarnate. This break expressed itself also in morality: the natural order as the basis for ethics is replaced by 'the law of love'. Finally, Christians believe that salvation does not imply a unification with the cosmic order but involves redemption and resurrection in a new body.

Ferry indicates that the emergence of modern philosophy involves a radical break with Christian thinking. Modern man becomes the foundation for understanding the order of the world, developing morality and realizing salvation. The idea of discovering the divine order is substituted by creating or constructing order as human beings. The law of love is replaced by a ratio-based ethics. Finally, any belief in a divine redemption is rejected in favour of salvation in the way of science and technology. Consequently, the idea of eternal life is rejected and it is believed that life ends with death.

The era of postmodern thinking has been ushered in by the philosopher Friedrich Nietzsche. Ferry demonstrates that there is a radical break between modern and postmodern thinking. Nietzsche believes that reality is not an ordered or harmonious unity, but an infinite multitude of forces and impulses that constantly collide. He believes that a universal ethics does not exist. Every individual human being has to develop his own values and his own 'grand lifestyle'. Finally, he thinks that our salvation lies in a life worth living, in an intense, exalted and courageous life. A life in which there is no room for regret and repentance.

What do we learn from the philosophical considerations of Luc Ferry? Our first conclusion is that every philosophical era is characterized by different beliefs about the order of reality, human relationships, and the meaning of life. The context in which we interpret our observations of society and nature is not a constant, and different belief systems make us draw different conclusions and take different actions. That also implies that, in doing scientific research and developing technological solutions for the energy transition we cannot ignore our beliefs about reality, fellow man, and the meaning of life. We have to make these beliefs explicit. Our second conclusion is that the course of western history can be described as a history of broken connections. The rise of modern thinking implied the breaking of the relationship of man with God. Additionally, the inherent and interdependent relationship of man and nature was broken and replaced by an instrumental one in which man exploits nature. The rise of postmodernity implied the breaking of the relationship of man with fellow man and perfected the break of human being and nature.

Latour states that modern man is disconnected from nature and fellow man. He also argues that these disconnections have a religious origin. Ferry gives relief to these observations. He showed that man was originally connected to the Logos or God, fellow man and nature (Greek thinking, Christian thinking). In the course of history, however, these connections have been broken (modern thinking, postmodern thinking). The breaking of these connections can be interpreted as an existential or religious act. As a result, restoring these connections also requires an existential or religious act.

The Question of the Sacred

Bronislaw Szerszynski also addresses the connections of man and nature. He describes these connections in terms of 'sacred' and 'secular'. At first sight, it seems to be farfetched to relate issues of the energy transition with words like 'sacred' and 'secular'. However, in the course of this section we will discover how important these concepts are.

Szerszynski wonders in his book *Nature, Technology and the Sacred* (2005), how we have to judge our time. He argues that Max Weber's vision of the 'disenchantment of the world' is widespread. Under the influence of science and technology we have stripped nature from mysterious powers and divine interventions. We no longer believe in gods, demons and spirits who can help, hinder or frighten us. We believe today that reality can be understood in mathematical terms and physical laws. In fact, we can control nature through science, realize its potential through technology, and determine its value in the market. In summary, 'disenchantment' means that 'religion has been replaced by science and technology' (Szerszynski, 2005, p. 14).

Szerszynski notes that present thinking is characterized by an asymmetrical vision of the sacred and the secular. The secular is seen as self-evident, as something that needs no explanation. But the sacred is interpreted as something aberrant, as something so special that it requires further explanation. It is precisely because of this asymmetry that he wants to 'problematize' the secular.

Szerszynski shows that the story of the concepts of the sacred and the secular is much more complex than described in many popular and philosophical reflections. In the classical world the term 'secular' or 'profane' has always been interpreted religiously. He refers to the original meaning of the word 'profane': *pro-fanum* is the space in front of the sanctuary. In other words, in classical thinking, the profane was always a space within a sacred cosmos. Modern thinking, however, states that the world is completely profane and has no spiritual meaning whatsoever. In this way of thinking the secular presents itself as a self-grounding, independent reality. Szerszynski wants to problematize this vision of reality. He believes that modern thinking has not disenchanted the world, but has replaced one belief by another. He even wonders if we should not see the story of the disenchantment of nature as the 'creation myth of modern society' (Szerszynski, 2005, p. 7). Szerszynski argues that modernity—and therefore modern views of the sacred and the secular—must be seen as a specific product of our religious and cultural history. He even calls the secular a religious phenomenon. In his view, the sacral in modern times has not disappeared. Rather, it has been ordered or organized in a different way.

Szerszynski uses the word 'sacred' in a general sense. He writes: "I am using 'sacred' in a more general sense, to understand the ways in which a range of religious frames are involved in our ideas of and dealings with nature and technology (…) it is the ground against which particular historical phenomena or ideas appear as intelligible figures" (Szerszynski, 2005, p. ix). He refers to the views of Kay Milton who defines the sacred as 'what matters most to people' and to the definition of Paul Tillich who describes religion as 'ultimate concern' (Szerszynski, 2005, p. ix).

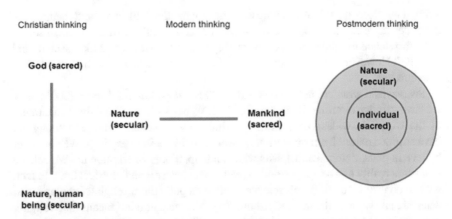

Fig. 2 The reordering of the sacred in the course of history

Szerszynski describes the development of the sacred under the heading 'The Long Arc of Transcendental Religion.' The story begins with the primal sacred of indigenous peoples who experience reality as a unity of the natural and the divine. They make no distinction between the empirical and the transcendent, the secular and the sacral. The story ends with the plurality of the postmodern sacred in which the unity of the natural and the divine has collapsed. The result is the emergence of a multiplex reality that is founded in the subjective experience of the individual.

We would like to highlight three stages in 'The Long Arc', see Fig. 2.

In the Protestant sacred the gap between the transcendent divine and the secular is seen as infinitely large and infinitely small, as absolute and disappearing into nothingness.[2] On the one hand, the Transcendent is depicted as the Exalted, the Almighty. On the other hand, He is near and directly accessible to the individual, without heavenly or earthly intermediary. The Protestant sacred opens the way for the individual—created in the image of God—to serve God in all areas of profane life. Szerszynski uses the word 'profane' almost in its original meaning here: the 'profane' is directly related to the sacred and acquires meaning from the sacred.

In the modern sacred, which encompasses Enlightenment and Romanticism, the vertical transcendent axis is increasingly being drawn into the empirical world. In the Protestant sacred, 'being' and 'order' in nature are related to a 'supernatural origin,' but in the modern sacred they are increasingly seen as properties of reality itself. The world is becoming profane in a new sense, namely, as a space that is only profane

---

[2]In Szerszynski's 'The Long Arc of Transcendental Religion' the Protestant sacred is preceded by the 'monotheistic sacred' of the historic religions, including the world religions of Judaism, Buddhism, Christianity and Islam. In his view, the monotheist sacred is characterized by a dualist distinction between this world and a transcendental reality. Given the line of thought we develop in this chapter, it is not necessary to discuss the monotheistic sacred.

and has no relationship with the sacred. The profane has become 'total' or 'absolute'. This development, however, does not lead to the disappearance of the sacred but to a reordering of the sacred: the sacrality of the human subject. In other words, man assigns a divine character to himself. The reordering of the sacred also leads to a new vision of salvation. Christ's redemption is replaced by self-redemption. In the tradition of Enlightenment, emphasis is placed on the path of science and technology and in the tradition of Romanticism on authenticity and solidarity with the world.

In the postmodern sacred, the Protestant sacred has collapsed entirely. A multiplex reality arises that is filled with and constituted by different views on man and reality that are founded on subjective experience. People no longer focus on a natural or divine order. In fact, they reject such an order that gives direction to their lives. Instead, they develop their own philosophy of life based on 'what feels right' and shape their 'own religion'. The idea of a common vision of man and reality, which was still present in the modern sacred, has given way to a plurality of visions.

What can we learn from Szerszynski's fascinating sketch of the history of the sacred and the secular? First, Szerszynski poses probing questions about our relationship with the earthly. In view of the climate crisis and the energy transition these questions are of an existential nature. These are questions like: What is still sacred for us? Or: What transcends our personal interests? Or: What may it cost us? Second, we conclude that Szerzynski—if his analysis is correct—sketches a gloomy picture. The sacred is concentrated in the individual. Every individual develops his or her own philosophy of life. In other words, there is no sacred that transcends the individual. If this is indeed the case, then our starting point for tackling overwhelming issues like the energy transition is not very favourable.

The Hidden Dimension in the Worlds of Engineers, Policy Makers, and Politicians

The message of Latour is that the hidden dimension is also present in the worlds of engineers, policy makers, and politicians. The philosophers Ferry and Szerszynski confirm this analysis in their ideas of broken connections and the absence of a shared meaning of the sacred. In this section we will investigate the question 'how' the hidden dimension is present in and penetrates the practices of engineers, policy makers, and politicians. We focus on these practices because they largely determine how humanity interacts with nature.

Practice Approaches

It goes without saying that the hidden dimension of the worlds of engineers, policy makers, and politicians cannot be investigated by scientific approaches that are based on objectivity, rationality, and progress. The main reason is that the scientific approach, according to Latour, in its core values denies the existence of this hidden dimension. Nicolini (2012) proposes that scientists have to change their research methods to understand what really happens in the worlds of engineers, policy makers, and politicians. In his view, scientists must not focus on positive facts and rational data, but on meanings, decisions and actions. De Vries and Jochemsen (2019) argue that these worlds have a normative character and are co-shaped by philosophical and religious ideas. The approaches advocated by Nicolini (2012) and De Vries and Jochemsen (2019) are called 'practice approaches' because they try to understand what really happens in the worlds of engineers, policy makers, and politicians. Moreover, they refer to 'social practices' to emphasize the social dimension of practice approaches in the worlds of engineers, policy makers, and politicians.

Practice approaches are fascinating as they deal with real world complexity and render insight into how practitioners perceive this complexity:

- *Complexity.* Practice approaches offer tools to characterize, analyse and understand the complexity of social practices.
- *Practice-centred.* Practice approaches take their starting point in social practices. They take the world of the engineer, policy maker, and politician seriously. They try to understand what drives them to excel and what values they strive to anchor in their work.
- *Bodies and technology.* Practice approaches bring to the fore that in all practices bodily activities and material things play a critical role. Human practices without body and technology do not exist.
- *Stakeholders.* Practice approaches recognize that social practices act in a complex world. In other words, every social practice has stakeholders that have an interest in the goings-on within a practice.
- *Spirit of the times.* Practice approaches are aware that there is something like 'spirits of the times'. The idea of the 'spirit of the time' is difficult to grasp. Despite that, practice approaches try to address this topic.
- *Human phenomena.* Practice approaches leave space for human phenomena like initiative, creativity, conflict, power, deceit, and so on. Practice approaches show that these types of phenomenon are relevant and co-determine the performance of practices.

In this chapter, we introduce the triple-I framework as a simple framework to help us understand the social practices of engineers, policy makers, and politicians. This practice framework was developed in close collaboration with engineers and managers, and provides us with a handle to explore where the hidden dimension is expressed in the social practices of engineers, policy makers and politicians.

Triple I Framework: Three Perspectives

The Triple I framework offers three different perspectives to investigate social practices (Verkerk, 2014, 2019). We would like to emphasize the word 'perspective'. The three 'I-s' do not stand for three different 'parts' of a practice but present three different points of view to understand social practices and the hidden dimension therein. Each one of the three perspectives reveals specific characteristics. Combining the perspectives results in a richer understanding of social practices. We distinguish the following perspectives:

(1) The first perspective is 'identity and intrinsic values'. This perspective focusses on the opinions and beliefs of the main actors in a social practice about their own role and identity and about the main values to be embedded in their work.

(2) The second perspective is 'interests of stakeholders'. This perspective highlights the interests of third parties (stakeholders) that have a stake in a social practice. It also underlines the mechanism by which these stakeholders exert influence on that practice.

(3) The third perspective is 'ideals and basic beliefs'. This perspective puts the spot light on the influence of ideals and basic beliefs in society on a social practice. It investigates how social practices are co-shaped by societal norms and values.

This framework is helpful to gain an understanding of the practice of engineers, the practice of policy makers, and the practice of politicians, see Fig. 3. Through a Triple I analysis of each of these practices a picture emerges of the similarities, differences, and complementarities of the three practices.

Identity and Intrinsic Values

The first I highlights the 'identity and intrinsic values' of a practice. It has to be noted that 'identity' and 'intrinsic values' are closely related. They can be described as two sides of a coin. On the one hand, the identity of a practice is specified in more detail by the intrinsic values. After all, it specifies which values are important in the practice concerned. On the other hand, the intrinsic values co-shape the identity of a practice. It goes without saying that the identities of the practices of engineers, policy makers, and politicians are quite different. The practice of engineers is about making technology work for society, the practice of policy makers is about designing effective and efficient policy interventions for the good of society (in health care, education, culture, industry, and so on), and the practice of politicians is about defining the 'good of society', i.e. solving or tackling value dilemmas, with respect for democratic legitimacy, and overseeing public administration for the manner in which it upholds societal values (fairness, equity, justice, transparency and so forth). Each of these practices have their own intrinsic values. For example, for engineers the values of creativity and innovation score high and for politicians the values of support and

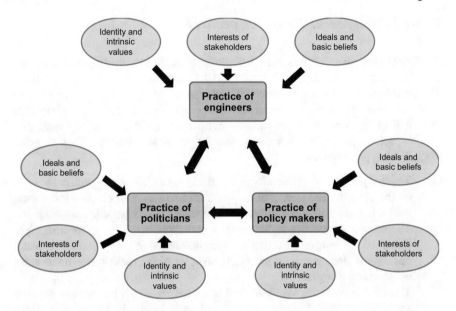

Fig. 3 The Triple-I framework for **a** engineering practices, **b** policy making practices, **c** and political practices. These practices can be understood from the perspectives 'identity and intrinsic values', 'interests of stakeholders', and 'ideals and basic beliefs'

feasibility. The Triple I framework suggests that hidden dimensions are present in the intrinsic values of each of these practices. The use of this framework invites engineers, policy makers and politicians to reflect on their intrinsic values and to wonder to what extent they express the connection between man and fellow man, and man and nature.

Interests of Stakeholders

The second I highlights the complex environments in which engineers, policy makers, and politicians operate. They are embedded in different social networks with many other actors and stakeholders. In other words, every practice has its own configuration of stakeholders. Among practices, these configurations may partly overlap. For example, engineers are mainly concerned with users, policymakers with citizens, and politicians with voters. Each stakeholder has an interest in influencing these practices to reach an outcome benefiting the relevant stakeholder. The Triple I framework suggests that hidden dimensions are present in the influence of the stakeholder configurations of each practice, and it invites engineers, policy makers and politicians to reflect on the influences of stakeholders on their practices. What types of values do they promote? Vales of 'externality', 'indifference', and 'detachment'? Or values of 'close proximity', 'involvement' and 'connectedness'? We do acknowledge that every stakeholder has its own *justified* interests. However, to what extent are their justified interests embedded in modern views on man, organization, and market that promote broken relationships?

Ideals and Basic Beliefs

The third I reveals the (hidden) ideals and basic beliefs in society. As said before, the practices of engineers, policy makers, and politicians cannot be seen as isolated from society at large. They are embedded in society as a whole, as all practitioners engaged in the different practices are also engaged in other social contexts, such as marriage, family, sports, leisure, church, and so on. In all these other social contexts, practitioners breathe in the ideals and basic beliefs of society, like the air they breathe, without realizing it. The framework suggests that the ideals and basic beliefs of the modern/postmodern society co-shape the hidden dimensions, as stipulated by Ferry and Szerzynski.

Feelings of Unease

Many engineers, policy makers, and politicians will argue that they act in an objective and rational way and that hidden dimensions do not pertain to their practice. They will also argue that stakeholders only have an 'objective' and 'rational' influence on their practice and that, in their work as practitioners, they are immune for ideals and basic beliefs in society. A recognition of hidden dimensions is certainly at odds with what practitioners perceive as good engineering practices and good policy making practices. Most practitioners have not been trained to be sensitive to hidden dimensions, nor to explicitly account for the influence of their world view, ideals and basic beliefs in their work. This is why we need to acquire a better understanding of how social practices are shaped and pervaded by hidden dimensions.

A Plea for the (Secular) Sacred

We would like to start with a short recap of our line of thought. The philosopher Bruno Latour suggests that the overwhelming dilemmas and the insolvability of the climate issue has to do with a 'hidden dimension' in our thinking. The first contour of this hidden dimension is related to the broken connections of man on the one hand and fellow man and nature on the other hand. The second contour is found in questions pertaining to the meaning of life that have a religious or philosophical background.

The philosopher Ferry sheds light on the idea of broken dimensions. He argues that broken connections do not just show up in our society but are the result of a long evolutionary process unfolding throughout the history of humankind. He shows that in this historical process society developed new ideas and beliefs about the meaning of life. These ideas and beliefs pervade society and social practices, without explicit recognition of their role in social practices. That is why they are referred to as a 'hidden dimension'. The philosopher Szerszynski approaches the 'hidden dimension' from the perspective of sacrality. In our postmodern society, he argues, every individual defines what is sacred to him or her in relationship to fellow man and nature. That means, there are no shared values in postmodern society to help us collectively address the climate issue.

The key question of Latour is: Is a radical revision of our relationship with the earthly possible? And what would be the contours of such a revision? Ferry rejects the idea that the present postmodern beliefs can be a source for a radical revision of the relationship of man with fellow man and nature. He defends a rehabilitation of the concept of transcendence (Ferry, 2010, pp. 232–239). The concept of transcendence implies that there is something that is greater that man, something that surpasses every individual and individual interests. He argues that values such as 'truth', 'beauty', 'justice' and 'love' do no originate from individuals but stem from how we experience and interact with others, in relationships between individuals and in social communities. The idea of transcendent values is very helpful to promote a radical revision of our relationship with the earth. Szerszynski also believes that the postmodern sense of the meaning of life prohibits us to tackle the challenges of our global society. He argues for a *concept of sacrality in which plurality and unity are connected to each other*. He searches for *an idea of a transcendent axis that makes a plurality of perspectives on reality possible* (Szerszynski, 2005, pp. 170, 175).

The philosophers Ferry and Szerszynski show us a way for man and fellow man, for man and nature to be reconnected. It is about *values that rise above us*. It is about the recognition that man is neither the origin of values, nor the creator of connectedness, nor the source of meaning, and nor the source of sacrality. On the contrary, it suggests that *being human has to do with the art of receiving: receiving values, receiving connections, receiving meaning, and receiving sacrality*. The *idea of receiving presupposes that there is 'somebody' or 'Somebody' who offers values, connections, meaning, and sacrality*.

What we can learn from the philosophers is that every practitioner, whether engineering professional, policy maker or politician, has a responsibility towards society and the planet to reflect on values, connections, meaning and sacrality. Inevitably, in a hyper-individualistic society like ours, practitioners' reflections on these deep questions will yield a large diversity of answers. The hard question then is: is there a common sacred or are there common values that connect man with fellow man and man with nature? We would like to point out that there are many national and international initiatives that transcend the plurality that characterizes our society. With regard to the international initiatives, we would like to draw attention to the Sustainable Development Goals as developed by the UN and the Paris (2015) UNFCCC climate agreements, which have again been signed by the United States of America. We would like to suggest three possible common values or forms of sacrality. The first one is the value or sacrality of the earth: mankind has no choice but to act in accordance with the rules of the earthly ecosystem; rules we learn through trial and error. The second one is the idea of the dignity of human beings: every human being has the right to live in a healthy and sustainable world. The third one is the idea that we should leave a good earth to our children and grandchildren. These three different values or sacralities do not exclude each other. Each of them recognizes the idea of receiving, the idea of something that transcends the individual, and the idea that there is something sacred that man should not enter or tarnish. Finally, each of them offers action perspectives.

Action Perspectives

In this chapter we have investigated the energy transition from a philosophical point of view, in the context of the climate change debate. Bruno Latour claims that the climate crisis has a 'hidden dimension'. He argued that the first contour of this dimension is found in the interpretation of the relationship of man with fellow man and the relationship of man with nature. In the course of history these connections have been broken. The second contour is found in the existential interpretation of man. It is about the idea that our ultimate beliefs about values, connections, meaning, and sacrality have a religious or philosophical origin. We have claimed that the relationship of man with fellow man and man and nature only can be reinstated by the recognition of values that transcend human being as an individual and/or the recognition of a (secular) sacred.

In our view, the Triple I framework helps us to identify potential action perspectives with respect to revealing the hidden dimension in the social practices of engineers, policy makers, and politicians. The first I (identity and intrinsic values) invites engineers, policy makers and politicians to address the identity and the hidden values in their own practices. In the context of the energy transition, the key question here is: do these values lead to a further increase in inequalities in society or to the notion that we have to restore broken relationships within society and between society and nature? The second I invites engineers, policy makers and politicians to engage in dialogue with stakeholders. These dialogues can contribute to a common understanding that there is no plan B for the planet. The third I invites practitioners to reflect on their religious values or their philosophical choices with respect to fellow man and our ecosystem.

We do not believe that technology 'as such' will solve the problems of the energy transition. We need religious, philosophical, and existential discussions about the human condition in times of climate crisis. Only through such discussions may we feel motivated, inspired and possibly morally obliged to realize the energy transition in an inclusive manner and unleash the creativity required to make this transition possible.

References

de Vries, M., & Jochemsen, H. (Eds.). (2019). *The normative nature of social practices and ethics in professional environments*. IGI Global.

Ferry, L. (2010). *Learning to live: A user's manual*. Canongate.

IPCC. (2013). *Climate change 2013: The physical science basis*.

Latour, B. (2017). *Facing Gaia: Eight lectures on the new climatic regime*. Polity Press.

Latour, B. (2018). *Down to earth: politics in the new climatic regime*. Polity Press.

Nicolini, D. (2012). *Practice theory, work, and organization: An introduction*. Oxford University Press

Szerszynski, B. (2005). *Nature, technology and the sacred*. Blackwell.

Verkerk, M. J. (2014). A philosophy-based "toolbox" for designing technology: The conceptual power of Dooyeweerdian philosophy. *Koers—Bulletin for Christian Scholarship, 79*(3), 1–7. Art. #2164. https://doi.org/10.4102/koers.v79i3.2164

Verkerk, M. J. (2019). Industrial practices, sustainable development and circular economy: Mitigation of reductionism and silo mentality in the industry. In Vries and Jochemsen.

The Technological Design Challenge

Hydrogen–The Bridge Between Africa and Europe

Ad van Wijk and Frank Wouters

Abstract This chapter describes a European energy system based on 50% renewable electricity and 50% green hydrogen, which can be achieved by 2050. The green hydrogen shall consist of hydrogen produced in Europe, complemented by hydrogen imports, especially from North Africa. Hydrogen import from North Africa will be beneficial for both Europe and North Africa. A bold energy sector strategy with an important infrastructure component is suggested, which differs from more traditional bottom-up sectoral strategies. This approach guarantees optimized use of (existing) infrastructure, has low risk and cost, improves Europe's energy security and supports European technology leadership. In North Africa it would foster economic development, boost export, create future-oriented jobs in a high-tech sector and support social stability.

Introduction

Electrification is one of the megatrends in the ongoing energy transition. Since 2011, the annual addition of renewable electricity capacity has outpaced the addition of coal, gas, oil and nuclear power plants combined, and this trend is continuing. Due to the recent exponential growth curve and associated cost reduction, solar and wind power in good locations are now often the least cost option, with production cost of bulk solar electricity in the sunbelt soon approaching the 1 $ct/kWh mark. However, electricity has limitations in industrial processes requiring high temperature heat, chemicals feedstock or in bulk and long-range transport.

Green hydrogen made from renewable electricity and water will play a crucial role in our decarbonized future economy, as shown in many recent scenarios. In a system soon dominated by variable renewables such as solar and wind, hydrogen links electricity with industrial heat, materials such as steel and fertilizer, space heating, and

A. van Wijk (✉)
Delft University of Technology, Delft, The Netherlands
e-mail: a.j.m.vanwijk@tudelft.nl

F. Wouters
Worley, Masdar City, Abu Dhabi, UAE

© The Author(s) 2021
M. P. C. Weijnen et al. (eds.), *Shaping an Inclusive Energy Transition*,
https://doi.org/10.1007/978-3-030-74586-8_5

transport fuels. Furthermore, hydrogen can be seasonally stored and transported cost-effectively over long distances, to a large extent using existing natural gas infrastructure. Green hydrogen in combination with green electricity has the potential to entirely replace hydrocarbons.

Due to its limited size and population density, Europe will not be able to produce all its renewable energy in Europe itself. Therefore, it is assumed that a large part of the hydrogen will be imported. Although hydrogen import can come from many areas in the world with good solar and wind resources, an interesting possibility is the import from North Africa. Already today, 13% of the natural gas and 10% of the oil consumed in Europe come from North Africa (Eurostatimports, 2019) and 60% of North Africa's oil exports and 80% of its gas exports are sent to Europe.

North Africa has good solar and wind resources and many countries are developing ambitious renewable energy strategies to cater for growing energy demand of urban and industrial centers, but also electrify the unserved parts of the population in remoter areas. Low-cost and price-stable renewable electricity has the potential to spur economic growth, necessary to stabilize societies and reduce economic migration. However, over and beyond catering for domestic demand, most North African countries have huge potential in terms of land and resources to produce green hydrogen from solar and wind for export. The resources in North Africa are vast. Only 8% of the Sahara Desert covered with solar panels suffice to produce all the energy for the world, 155,000 TWh per year (Wijk et al., 2017).

If Europe and North Africa can develop a joint hydrogen economy, both North Africa and Europe will benefit. Only the Mediterranean Sea separates the two regions. Hydrogen can be imported from North Africa by pipeline, which is more cost effective than import by ship. With hydrogen imported from North Africa, Europe could realize a sustainable energy system, required to meet the obligations of the Paris Agreement, faster and cheaper. Furthermore, a joint European—North African renewable energy and hydrogen approach would create economic development, future-oriented jobs and social stability in North-African countries, potentially reducing the number of economic migrants from the region to Europe.

Renewable Energy Resources in Europe and North Africa

In Europe, good renewable energy resources are geographically distributed. However, they are not evenly distributed among the EU member states and therefore large scale, pan-European energy transport and storage is necessary.

Large scale on- and offshore wind can be produced at competitive and subsidy free prices in several parts of Europe. Large scale offshore wind has great potential in the North Sea, Irish Sea, Baltic Sea and parts of the Mediterranean Sea. And large-scale onshore wind potential can be found especially in Greece, the UK, Ireland and in many other coastal areas in Europe such as Portugal, Poland and Germany. Large scale solar PV can also be built competitively and subsidy-free, most notably in Southern Europe, for instance in Spain, Italy and Greece.

Furthermore, low cost hydropower electricity can be produced in Iceland, Norway, Sweden, Austria, Switzerland, etc. and geothermal electricity in Iceland, Italy, Poland and Hungary. Although, the potential expansion of the hydropower and geothermal capacity is limited, the future introduction of marine/tidal energy converters could furthermore augment the production of renewable electricity and hydrogen in the UK, Portugal, Norway and Iceland.

In North Africa, however, the solar energy resources are even better than in Southern Europe. The Sahara Desert is the world's sunniest area year-round. It is a large area (more than 9.2 million square kilometer) that receives, on average, 3600 h of sunshine yearly and in some areas 4000 h. This translates into solar insolation levels of 2500–3000 kWh per square meter per year (Varadi et al., 2018). A fraction of the Sahara Desert's area could generate the globe's entire electrical demand.

Also, it should be noted that the Sahara Desert is one of the windiest areas on the planet, especially on the west coast. Average annual wind speeds at ground level exceed 5 m/s in most of the desert and reach 8–9 m/s in the western coastal regions. Wind speeds also increase with height above the ground, and the Sahara winds are quite steady throughout the year (Varadi et al., 2018). Also, Egypt's Zaafarana region is comparable to Morocco's Atlantic coast, with high and steady wind speeds, critical for the economics of wind energy as the energy derived from a wind turbine scales at the third power of the speed of the air passing through its blades. In Morocco, Algeria and Egypt certain land areas have wind speeds that are comparable to offshore conditions in the Mediterranean, Baltic Sea and some parts of the North Sea.

One should consider the difference between countries that are net energy importers, such as Tunisia and Morocco, and net exporters such as Algeria, Libya and Egypt. Morocco has been leading the pack and is embarking on an ambitious renewable journey, building world class solar and wind projects to increase energy security, reduce cost, emissions and price volatility and support economic growth. For Algeria and Egypt, tapping into low-cost renewables reduces overall system cost and frees up fossil fuels for higher value applications and export.

Large scale solar PV, Concentrating Solar Power and wind can be realized in North African countries against production cost lower than in Europe. The expectation is that solar PV and wind onshore production cost in North Africa will come down to 1 $ct/kWh before 2030 (Fig. 1).

Energy in Europe

Energy carriers are used for heating, mobility, electricity and in industry for high temperature heat and as a feedstock. In 2017, the total energy consumption (Gross Available Energy) in the European Union amounted to 1719 Mtoe or almost 20,000 TWh (EurostatEnergy, 2019). Final energy consumption in 2017, the energy consumed by end users, was 1123 Mtoe or about 13,000 TWh, see Table 2. The European Union is a net energy importer, with 55% of the 2017 energy needs (Gross Available Energy) met by imports, consisting of oil and oil products, natural gas and

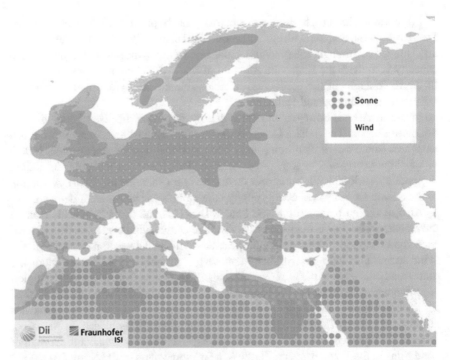

Fig. 1 Solar irradiation and wind speed in Europe and North Africa. North Africa has world class solar and wind resources (Dii & Fraunhofer ISI, 2012)

solid fuels. Although Europe is working ambitiously to become less dependent on energy imports, it is unlikely that it can become entirely energy self-sufficient. Most scenarios, including BP's Energy Outlook 2019 (BP, 2019) indicate that Europe shall remain a net importer of energy until mid-century and beyond. Given the population density and comparatively limited potential for renewable energy, the expectation is that Europe shall continue to import energy, also in a future renewable energy system. However, instead of fossil fuels, over time Europe shall import energy in the form of green electrons, but especially in the form of green molecules (Table 1).

To meet their obligations under the Paris Agreement, the EU Member States have set key targets for 2030: (1) a 40% cut in greenhouse gas emissions compared to 1990 levels, (2) at least a 32% share of renewable energy consumption and (3) an improvement in energy efficiency at EU level of at least 32.5%. The corresponding EU 2030 goal for final energy consumption is set at 11,118 TWh (Eurostat, 2019).

Beyond that, the European Commission calls for a climate-neutral Europe by 2050, laid down in the document "A Clean Planet for all", which was released in November 2018 (European Commission, 2018). The fuel mix in Gross Inland Consumption, that is projected in 2050, under different scenarios, is shown in Fig. 2. In all these 2050 scenarios fossil fuels and nuclear still have a significant share. There is a current debate ongoing about which scenario is most appropriate for

Table 1 EU28 2017 energy consumption (EurostatEnergy, 2019)

2017 EU28	Mtoe	TWh
Gross available energy	1,719	19,993
- International maritime bunkers	-45	-523
Gross inland consumption	1,663[a]	19,341
- Feedstock	-102	-1,191
Primary energy consumption	1,561	18,150
- Conversion losses energy sector	-438	-5,094
Final energy consumption	1,123	13,056

[a] Ambient heat (11 Mtoe) is also subtracted (EurostatGuide, 2019)

Table 2 EU28 2017 Final energy consumption (Eurostat Energy, 2019)

2017 EU28	TWh	%
Oil	4,584	35
Gas	2,783	21
Electricity	2,798	21
Renewables + Biofuels	1,190	9
Solid fuels	298	2
Other	1,404	11
Final energy consumption	13,056	100

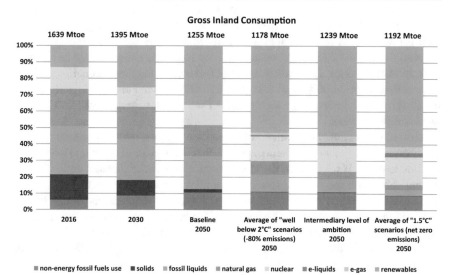

Fig. 2 Fuel mix for gross inland consumption EU28, projected for 2050, for different scenarios from the EU document 'A clean planet for all' (European Commission, 2018)

Table 3 Solar and wind energy in the European union in 2050, according to several scenarios

Scenario	Solar energy [TWh/a]	Wind energy [TWh/a]	Solar capacity [GW]	Wind capacity [GW]
Shell sky scenario	3,472	3,089	2,300	1,000
DNV GL energy transition outlook 2018	1,077	1,662	718	554
LUT/EWG			2,000	560

Europe, with several European member states arguing that Europe needs to pursue a 100% renewable energy scenario (Morgan, 2019).

Several recent scenarios exist for Europe's energy system in 2050, including Shell's Sky Scenario (Shell, 2018), The Hydrogen Roadmap for Europe (FCHJU 2019), DNV-GL's Energy Transition Outlook 2018 (DNV-GL 2018) and the "Global Energy System based on 100% Renewable Energy–Power Sector" by the Lappeenranta University of Technology (LUT) and the Energy Watch Group (EWG) (Ram et al., 2017). The following table contains a summary of the most ambitious renewable energy shares in each of these modeling exercises (Table 3).

It should be noted that to achieve the binding Paris Agreement, Europe's electricity sector needs to be fully decarbonized by 2050 and other energy sectors to a large extent also. This is a prerequisite for the Shell, GWEC and LUT/EWG scenarios. However, the DNV-GL ETO scenario is not compatible with keeping global warming well below 2 °C. It is reasonable to assume that for the DNV-GL scenario to be compatible with the Paris Agreement, the amount of solar energy would be closer to the results of the other scenarios. Analyzing and comparing these scenarios, one can assume that some 2,000 GW of solar and 650 GW of wind energy capacity can be installed by 2050, generating roughly 2,800 TWh of solar energy and 2,000 TWh of wind energy per year.

Most scenarios consider a drawn-out transition process, with a continuing dependency on fossil fuels, most of them imported, that will last for decades and would lead to climate chaos if released in the atmosphere. Since the associated emissions are incompatible with the Paris Agreement, several scenarios therefore feature massive investments in carbon capture and storage as well as future carbon sinks, mostly achieved through forestation. The Shell Sky scenario for example, contains a staggering 10,000 CCS projects necessary to limit CO_2 emissions. As of 2019, there are 21 CCS projects in the world (Carbon Capture and Storage, 2019) and less than 7000 coal fired power plants, so it would require a huge effort, technically, financially as well as regarding popular sentiment, to realize this many CCS projects. The question is whether there are no better alternatives altogether.

Energy in North Africa

The Southern Mediterranean countries can be currently divided into net energy importing and net energy exporting countries. Libya and Algeria have built their economies on the back of their substantial oil and gas reserves, whilst Morocco has always had to import fossil fuels. Egypt's recent offshore gas finds are expected to make the country a net natural gas exporter, joining Algeria and Libya. In the African context, in North Africa less than 2% of the population is without access to electricity. In contrast, 50% of people in West Africa and 75% in East Africa lack access to electricity. North Africa on average consumes eight times more electricity per capita than the rest of the continent, excluding South Africa (IRENA, 2015).

IRENA's Renewable Energy roadmap for Africa 2030 (IRENA, 2015) has analyzed options for the doubling of renewable energy supply by 2030 in a bottom-up approach. Supported by the excellent solar and wind resources in North Africa, it showed a feasible expansion to almost 120 GW by 2030, of which 70 GW would be wind and the remainder a combination of CSP (Concentrating Solar Power) and PV.

Morocco is an interesting example, as they have embarked on an ambitious renewable energy program with a target of 42% of renewable electricity by 2020. The state-owned entity MASEN plays a pivotal role. MASEN pre-develops renewable energy sites, carries through the procurement process, acts as the government entity borrowing concessional finance from development finance institutions and commercial lenders, and co-invests on behalf of the government. In Ouarzazate, a city in the south of Morocco's High Atlas Mountains they have built the Noor solar complex, consisting of CSP and PV projects, totaling 582 MW at peak when finished. The scale of these projects and Morocco's clever financial engineering have brought down the cost of CSP, which is now competitive with conventional power.

Hydrogen in Europe and North-Africa

Green hydrogen can be produced in electrolyzers using renewable electricity, can be transported using the natural gas grid and can be stored in salt caverns and depleted gas fields to cater for seasonal mismatches in supply and demand of energy (HyUnder, 2013). Like with natural gas, underground storage would be seasonal, while line-packing flexibility provides some short-term storage.

It should be noted that blue hydrogen, hydrogen produced from fossil fuels and combined with CCS, can play an important role in an intermediate period, helping kickstart hydrogen as an energy carrier alongside the introduction of green hydrogen.

Production Cost of Hydrogen

Renewable electricity is rapidly becoming cheaper than conventional electricity made in nuclear, gas- or coal-fired power plants. Already to date, solar power in Southern Europe and offshore wind in the North Sea does not require subsidy but can be sold at market prices. In North Africa, however, the electricity production costs with solar and wind are even lower than in Europe.

Green hydrogen is currently not cost-competitive compared to hydrogen made from hydrocarbons. Although for every ton of hydrogen produced today using steam methane reforming some 10 tons of CO_2 are released in the atmosphere, the price of carbon is not reflective of the cost to the global economy. There is no market yet for green hydrogen and electrolyzer manufacturers lack scale, resulting in relatively high cost of equipment. However, if a market would develop, hydrogen can be produced on locations with good solar or wind resources at € 1 per kg.

In January 2019, Morocco announced bids of € 28 per MWh for an 850 MW wind farm. The expectation is that electricity production cost will further drop to € 10–20 per MWh before 2030 at sites with good solar and wind resources throughout North Africa. Combinations of solar and wind, or even wind alone, will have load factors of 4,000–5,000 hours per year. With electrolyzer efficiencies of 80% (HHV, higher heating value) and CAPEX of € 300 per kW, the levelized cost of hydrogen production will be about € 1 per kg, see Fig. 3.

In Europe, however, with higher electricity production cost for solar and wind than in North Africa, the hydrogen production cost is expected to be € 0.5–1.0 per kg higher than in North Africa by 2030. But in 2050, with lower electricity production cost, higher electrolyzer efficiencies and lower CAPEX the hydrogen production cost

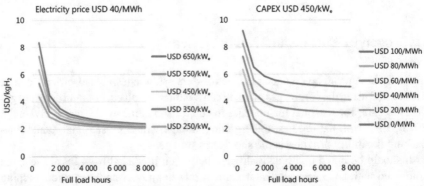

Notes: MWh = megawatt hour. Based on an electrolyser efficiency of 69% (LHV) and a discount rate of 8%.

Fig. 3 Future levelized cost of hydrogen production by operating hour for different electrolyzer investment costs (left) and electricity costs (right), from the future of hydrogen (IEA, 2019). (LHV Lower Heating value of hydrogen is 120 MJ/kg. HHV Higher Heating Value of hydrogen is 141.7 MJ/kg. An efficiency of 69% on LHV is equal to an efficiency of 81% on HHV)

will come down to € 1 per kg in Europe too. However, the production cost in North Africa, in 2050 will also drop and be well below € 1 per kg.

Infrastructure in Europe

In Europe, the lowest cost renewable resources are hydropower in Norway and the Alps, offshore wind in the North Sea and the Baltic Sea, onshore wind in selected European areas, whereby the best solar resource is in Southern Europe. The current electricity grid was not built for this, is not fit for the energy transition and needs to be drastically modernized. In 2018, an estimated € 1 billion worth of offshore wind energy was curtailed in Germany due to insufficient transmission grid capacity, according to the German Federal Network Agency (Bundesnetzagentur, 2019). In addition, the development of new renewable energy capacity is slowed down due to the lack of grid capacity. Unfortunately, overhead power lines are difficult to realize due to environmental concerns, popular opposition and typically take more than a decade for planning, permitting and construction.

However, a gas grid is much more cost-effective than an electricity grid: for the same investment a gas pipe can transport 10–20 times more energy than an electricity cable. Also, Europe has a well-developed gas grid that can be converted to accommodate hydrogen at minimal cost. Recent studies carried out by DNV-GL (2017) and KIWA (2018) in the Netherlands concluded that the existing gas transmission and distribution infrastructure is suitable for hydrogen with minimal or no modifications. So instead of transporting bulk electricity throughout Europe, a more cost-efficient way would be to transport green hydrogen and have a dual electricity and hydrogen distribution system. Figure 4 shows the existing European natural gas grid (blue) and a hydrogen backbone (orange) as suggested by the 2×40 GW Green Hydrogen Initiative, Hydrogen Europe (Wijk & Chatzimarkakis, 2020). Such a hydrogen backbone would link the areas of low-cost renewable electricity with the load centers in Europe. Operational by 2030–2035, it could be the first phase to realize a full conversion from natural gas to hydrogen by 2050.

The cost to build new hydrogen pipelines is comparable with the cost to build new natural gas pipelines. Europe also has an extensive network of offshore gas pipelines, an example of which is the Nordstream pipeline between Russia and Germany, 1,224 km long, with an investment cost of 7.4 billion Euro and a design capacity of 55 bcm/annum, or 68 GW (Nordstream, 2017). The Nordstream pipeline consist of 2 pipes with a diameter of 48 inch each. This pipeline was commissioned in October 2012 and has been upgraded over the years. In 2018, 58.8 bcm or 630 TWh natural gas was transported by Nordstream to Europe. An analysis of the investment cost for several of these large-scale pipelines shows an average turnkey investment cost of 1 million Euro per 10 GW per km.

Fig. 4 Natural gas infrastructure in Europe (blue and red lines) and first outline for a hydrogen backbone infrastructure (orange lines). The main part of the hydrogen backbone infrastructure consists of re-used natural gas transport pipelines with new compressors. A new pipeline from the solar and wind resource areas in Greece needs to be realized

Infrastructure Europe–North Africa

The electricity grid infrastructure in North Africa is not well developed, requiring major reinforcements and expansion in the coming decades, especially to transport electricity from the good solar and wind resource areas to the demand centers in the cities and rural areas.

Today, there are only two electricity grid connections between Europe and North Africa, each 700 MW grid interconnectors between Spain and Morocco. In the beginning of the century, the Desertec vision proposed to produce large amounts of solar and wind electricity in North Africa and expand the interconnection between Africa and Europe, enabling the export of part of this electricity to Europe across the Mediterranean. The cost to build such an electricity grid was huge, so even with lower production cost in North Africa, it was difficult for the imported electricity to compete with solar and wind electricity produced in Europe.

However, there is a gas transport infrastructure available between North Africa and Europe, transporting gas from Algeria and Libya to Europe via Italy and Spain. The gas transport volume through these pipelines is over 63.5 bcm per year, which equals a capacity of more than 60 GW (Timmerberg & Kaltschmitt, 2019).

Fig. 5 Natural gas infrastructure Europe–North Africa (left figure) and first outline for a hydrogen backbone infrastructure Europe–North Africa (figure above) An existing gas infrastructure from Algeria and Morocco could be converted to a hydrogen infrastructure (grey-orange lines). A "new" hydrogen transport pipeline must be realized from Italy to Greece, crossing the Mediterranean Sea to Egypt, which could eventually be extended to the Middle East (orange line)

In a first phase, between 2030 and 2035, the natural gas infrastructure could be used to transport hydrogen from North Africa to Europe. In an initial phase, a substantial hydrogen volume can be produced by converting natural gas to hydrogen, whereby the CO_2 is stored in empty gas/oil fields (blue hydrogen). Over the years however, with ever declining cost of renewable electricity and electrolyzers, more and more green hydrogen from solar and wind electricity can be fed into these pipelines.

Next to converting existing pipeline infrastructure, new hydrogen gas pipeline infrastructure could be built, connecting the good solar and wind resources in North Africa to Europe. A first new pipeline could be realized to connect Egypt and Greece to the main European gas grid in Italy, see Fig. 5.

The realization of a hydrogen "South–Nordstream" from Egypt, via Greece to Italy, 2,500 km, with a similar capacity as the actual Nordstream, with 66 GW capacity, consisting of 2 pipelines of 48 inch each, would imply total investments of € 16.5 billion. The cost figures are derived from the Nordstream project (Nordstream, 2014) and a study for US DOE (James et al., 2018). With a load factor of 4,500 h per year, an amount of 300 TWh or 7.6-million-ton hydrogen per year can be transported. Given the assumptions, as shown in Table 4, the levelized cost for hydrogen transport by pipeline, is 0.005 €/kWh or 0.2 €/kg H2, which is a reasonable fraction of the total cost of delivered hydrogen.

Table 4 The energy transport volumes and levelized cost of hydrogen transport for a "South–Nordstream", connecting the good solar and wind resources in Egypt and Greece to the European gas grid in Italy

South–Nordstream: Egypt–Greece–Italy Levelized cost of hydrogen transport by pipeline		
Assumptions		
Pipeline diameter	inch	48
Number of pipelines		2
Pipeline pressure	bar	100
Pipeline flow speed	m/s	30
Pipeline capacity	GW	2*33 = 66
Pipeline length	km	2,500
Specific investment cost	€/10 GW/km	1,000,000
Capex (Total investment cost)	Billion €	16.5
O&M cost (including compressor energy)	% Capex/yr	1
WACC (Weighted average cost of capital)	%	7
Lifetime	yr	40
Load factor pipeline	hr/yr	4,500
Calculations		
Energy transport	TWh/yr	300
	Ton H_2/yr	7.6 million
Levelized cost of hydrogen transport	€/kWh	0.005
	€/kg	0.2

Hydrogen Storage

Energy supply and demand always need to be balanced. Balancing oil and coal supply and demand is relatively easy and cheap by storing oil in tanks and coal in bunkers or in the open air. However, balancing electricity and gas supply and demand is more challenging.

Electricity supply and demand needs to be balanced at any moment in time. Balancing the electricity system today is mainly done using pumped hydropower and by flexible power plants, especially gas fired power plants. Natural gas storage is therefore crucial today in balancing electricity supply and demand. But an even larger seasonal gas storage volume is needed to balance gas production and supply for space heating.

Natural gas demand in Europe, especially in Northern Europe, shows a strong seasonal variation, in wintertime, the gas demand is 2–3 times higher than in summertime. However, natural gas production is constant throughout the year. Therefore, large scale seasonal storage of natural gas is necessary. Natural gas is stored in empty gas fields, porous rock formations and salt caverns. The overall storage capacity in

operation within the EU amount to 89.2 bcm (871 TWh). The largest storage capacities for gas are in Germany, in total 21.8 bcm (213 TWh). Half of this storage is
in salt caverns, amounting to 10.9 bcm (106.5 TWh) (Timmerberg & Kaltschmitt,
2019). Germany has by far the largest gas storage capacity in salt caverns in Europe,
but they are in use in several other countries too, see Fig. 7.

Total gas consumption in the EU in 2017 was 493 bcm (5,163 TWh) (Eurostatgas,
2019).

This total gas consumption includes 2,782 TWh as final gas consumption, with the
remainder used as feedstock and for electricity production. The storage capacity is
therefore 18% of total gas consumption in Europe. This storage capacity is especially
necessary to balance large scale, seasonal, weekly and daily gas demand fluctuations
especially for space heating and to a lesser extent for electricity. Part of this storage
capacity is in use to store energy for strategic reserves.

In a future energy system, the share of electricity from variable sources such as
solar and wind in the overall energy supply will dramatically increase. Although
the share of electricity is expected to grow to 50% of all final energy demand, green
molecules will be necessary for applications that are difficult or expensive to electrify.
Due to the variability of renewable energy sources and the large fluctuations in energy
demand for space heating and electricity, storage capacity is needed on an hourly,
daily, weekly and seasonal scale. Capacitors and batteries will play a significant role
for hourly and daily storage. For large scale, seasonal and weekly storage, hydrogen
storage, replacing natural gas storage, will become crucial.

Several studies have examined the need for hydrogen in an electricity system that
is increasingly based on renewable energy sources. Figure 6 shows how the need
for hydrogen grows exponentially in a system with variable electricity sources, as

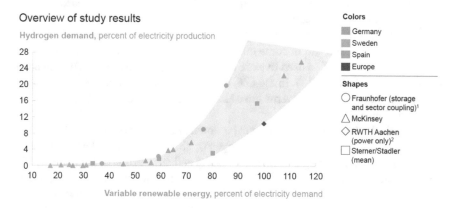

1 Least-cost modeling to achieve 2°C scenario in Germany in 2050 in hour-by-hour simulation of power generation and demand; assumptions: no regional
distribution issues (would increase hydrogen pathway), no change in energy imports and exports
2 Simulation of storage requirements for 100% European RES; only power-sector storage considered (lower bound for hydrogen pathway)

SOURCE: Fraunhofer Institute for Solar Energy Systems ISE, 2017; BMW; RWTH Aachen; Sterner and Stadler (2014); McKinsey

Fig. 6 The need for hydrogen in the electricity system increases exponentially with increasing
renewable energy share (Hydrogen council, 2017)

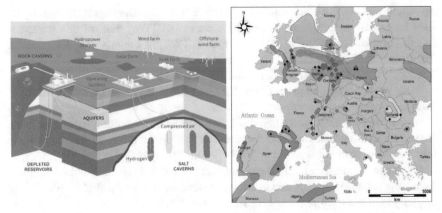

Fig. 7 Salt cavern (van Wijk, van der Roest and Boere 2017) (left) and salt formations with salt caverns throughout Europe (right). The red diamonds are salt caverns in use for natural gas storage (Bünger et al., 2016)

modeled by several institutions. When electricity systems are fully based on renewable energy sources, some 20% of variable electricity must be converted to hydrogen to guarantee a secure energy supply every time of the day and year.

The need for cheap hydrogen storage will grow exponentially over time. Salt caverns can provide this cheap hydrogen storage solution. Europe has still many empty salt caverns available for large scale hydrogen storage, but dedicated salt caverns for hydrogen storage capacity can be developed in the different salt formations in Europe. Potentially, hydrogen can be stored in empty gas fields that meet specific requirements to store hydrogen. However, this needs more research.

Salt caverns today are 'left over' from salt production. A typical salt cavern has a height of 300 m and a diameter of 60–70 m. A number of these salt caverns are in use for natural gas storage and in some other oil, compressed air or other products are stored. Salt caverns can be used to store hydrogen in the same way as they can store natural gas (HyUnder, 2013). In the UK, a salt cavern has been in use for hydrogen storage and in the US, salt caverns have also been used to store hydrogen for many years (see Fig. 7).

In a typical salt cavern, hydrogen can be stored at a pressure of about 200 bar. The storage capacity is about 6,000 ton hydrogen or about 240 GWh (HHV). The total installation costs, including piping, compressors and gas treatment, are about € 100 million (Michalski et al., 2017). For comparison, if this amount of energy would be stored in batteries, with costs of 100 €/kWh, the total investment cost would be € 24 billion.

In a recent study by Jülich research center (Caglayan et al., 2020), the potential for hydrogen storage capacity in salt caverns, that are especially leached for hydrogen storage, was investigated (see Fig. 8). There is a huge potential for hydrogen storage in salt caverns all over Europe. Total onshore salt cavern storage capacity is 23,200 TWh of which 7,300 TWh could be developed taking into account a maximum distance

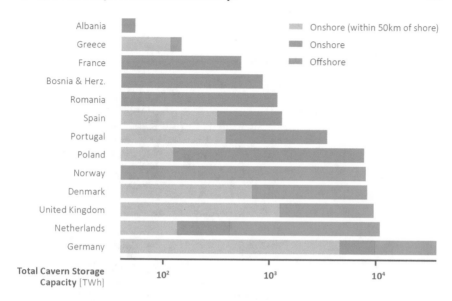

Fig. 8 The potential hydrogen storage capacity in salt caverns in Europe (Caglayan et al., 2020)

to the shore of 50 km, called the constrained storage capacity. This maximum limit is set for the brine disposal. The offshore storage capacity is even larger than the onshore capacity, 61,800 TWh. It should be noted that the salt cavern storage capacity potentials are even larger than total final energy consumption in Europe. Although not studied so far, a substantial potential for hydrogen storage in salt caverns is available in North Africa too.

A Different Approach

By 2050 when Europe's energy system is largely based on variable renewables, hydrogen is indispensable for transport and storage. Electricity demand will increase up to 2050, but there is a need for green molecules too. And, in an electricity system based on renewable energy resources, the need for hydrogen for storage and providing balancing power is evident.

The shares of hydrogen, presented in recent scenarios are by no means the maximum levels, nor do they represent the optimum. Several scenarios have tried to estimate the increasing demand for green hydrogen in Europe over time, most recently in the Hydrogen Roadmap by the Fuel Cells and Hydrogen Joint Undertaking (FCHJU 2019). This Roadmap estimates that hydrogen could comprise 2,250 TWh or 24% of Europe's total final energy demand including feedstock in 2050, using a gradual phasing-in approach, see Fig. 9:

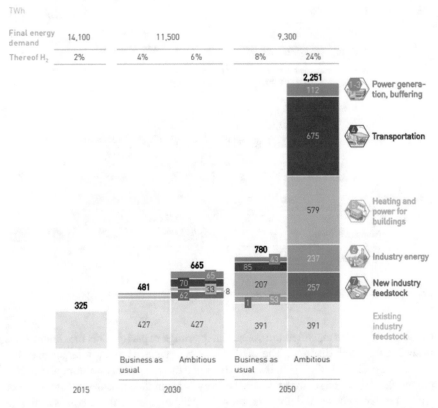

Fig. 9 Final energy demand including feedstock in the EU28 and the share of hydrogen demand for a business as usual and ambitious scenario, from the Hydrogen Roadmap (FCHJU, 2019)

Such scenarios typically use a bottom-up approach in a consultative process, analyzing various end-use sectors such as transport, the built environment, industry and the energy sector. Although there is merit in this approach by applying industry's collective knowledge and a deep-dive in these sectors, the fundamental flaw lies in the fact that at present there is no market for green hydrogen, and it is therefore very difficult to estimate e.g., adoption rates in industry, for fuel cell vehicles or the willingness among consumers to choose between green gas or all-electric solutions for their domestic energy needs.

Therefore, a different approach is proposed to realize a sustainable and inclusive energy system. A sustainable and inclusive energy system that is reliable, secure, affordable, accessible and fair. But above all, could be realized cheaper and faster than presented in recent scenarios by the EU and others. This approach could therefore offer a more realistic pathway for a climate-neutral and fully renewable energy system in the European Union by 2050.

1. Re-use gas infrastructure
2. Develop hydrogen storage

3. 50-50-50; Final energy demand in 2050, 50% electricity and 50% hydrogen
4. Europe needs North Africa for green hydrogen.

Re-Use Gas Infrastructure

Discussing the role of infrastructure in the energy transition, it may be more appropriate and insightful to look back and learn from the introduction of electricity and natural gas 100 and 50 years ago. Electricity and gas grids were built by governments and the service was offered to consumers, who rapidly stopped burning coal and candles in their houses and adopted these new energy sources in their industrial processes.

Following this analogy, instead of a gradual phasing in of green hydrogen, a more ambitious approach based on infrastructure development is therefore proposed. The fundamental philosophy is to make green hydrogen available at scale and cost-effectively and replace fossil fuels as quickly as possible by repurposing the current natural gas infrastructure to carry green hydrogen. Since the transmission and distribution infrastructure is already to a large extent available, the focus can be on developing electrolyzer capacity, which is an opportunity for European market leadership. Hydrogen's intrinsic quality as a transport fuel, its ubiquitous characteristics in industrial processes and ability for storage and long-range transport will lead to a rapid market uptake in Europe.

Initially a combination of blue and green hydrogen would be required to produce enough volume to convert a meaningful part of the European gas transport infrastructure. Over time blue hydrogen would be phased out and replaced by green hydrogen. At the latest in 2050 blue hydrogen will be fully phased out.

Of course, the electricity grid needs to be expanded and drastically modernized too, when a 50% share of electricity in final energy use is foreseen. However, the capacity in the electricity transport grid today is about 10 times less than the capacity in the gas transport grid. Besides, the cost for converting the gas grid to hydrogen will be much less, most probably a factor of 100 less, than building new electricity transmission grid capacity. And the natural gas grid, to a large extent, already exists, facilitating a much faster integration of renewable resources. A smart combination of expanding the electricity grid and at the same time re-using and expanding the gas grid for hydrogen will contribute to a cheaper energy infrastructure, at the same time realizing a renewable energy system faster.

Develop Large Scale Hydrogen Storage

In a future renewable energy system, largely based on variable renewable energy production and considering large scale fluctuations in daily, weekly and seasonal energy demand, large scale energy storage is needed. Today natural gas storage plays that vital role, with gas stored in empty gas fields, porous rock formations and salt caverns. Hydrogen can take over that role, especially by storing hydrogen in salt caverns, both existing and newly excavated. Storage capacity for natural gas comprises 18% of total gas consumption in the EU. In 2050, when the energy system is based on renewable energy sources, an assumed storage capacity of about 20–30% of final energy consumption is needed. Salt caverns can provide enough hydrogen storage capacity for this, catering for seasonal storage but also to keep a strategic energy reserve.

Large scale, seasonal and strategic energy storage can be provided by salt caverns relatively cheaply. However, battery storage will provide shorter term storage; hourly and daily storage and frequency control services for the electricity system can be provided by electrochemical batteries. But also, smart grids, demand side management, strengthening interconnections and other balancing instruments will be necessary to operate the electricity system reliably and cost-efficiently.

50–50; 2050 Final Energy Demand Split in 50% Electricity and 50% Hydrogen

Europe's final energy demand including feedstock and energy for international transport (international shipment and air-traffic), is estimated to be 12,000 TWh by 2050, based on an analysis of above-mentioned scenarios. If a similar division in energy use between the sectors is assumed as in 2017 (Eurostat, 2017), the final energy use per sector for 2050 is as shown in Table 5.

Today, electricity comprises less than 20% of final energy demand, including feedstock and international transport. About 80% of final energy demand are molecules, mainly fossil fuels. A small percentage is heat. The share of electricity in final energy demand is expected to grow 2.5 times until 2050, still leaving a large requirement for green molecules across all sectors. There are few alternatives to hydrogen, if any, to fully replace hydrocarbons in a decarbonized energy system, so electricity and hydrogen will both play an important role as energy carriers in the 2050 energy

Table 5 Share of EU final energy use per sector

Sector	TWh/a (2050)	Share (2017) (%)
Industry energy	2,500	21
Industry feedstock	1,300	11
Transport in EU	3,100	26
Transport international	700	6
Commercial and services	1,500	12
Households	2,700	22
Other	200	2
OVERALL	12,000	100

system. Therefore, a 50–50% share split of green electricity and green hydrogen in Europe's final energy demand is proposed for all sectors: industry, transport, commercial and households.

Of course, this is a rough estimate and will differ per sector and country. But it is doable in the transport sector, achieving a balanced mix of battery electric mobility for shorter distances, combined with fuel cell vehicles for heavy duty, longer ranges and higher convenience. In international transport and industrial feedstock, hydrogen will most probably have a bigger share than 50%. In these two sectors, there will be also a need for carbon (CO or CO_2), to produce chemicals and synthetic fuels, which will originate from biomass or by re-circulating the carbon from waste products. Most industrial high heat demand, currently served by natural gas, can be provided by hydrogen, and the household sector will consist of a mix of all-electric well-insulated new houses, while a large part of the existing building stock can be heated using hydrogen fuel cells and hydrogen gas boilers. Where the resource is available, district heating systems using geothermal or waste heat will play a role. Interesting future solutions also include the combination of heat pumps and hydrogen gas boilers, or hybrid geothermal heat pumps with fuel cells, in which the hydrogen boiler or fuel cell is responsible for the peak demand in the winter season.

Europe Needs North Africa for Green Hydrogen

Given a final energy demand of 12,000 TWh in 2050, with a 50–50% split between electricity and hydrogen, the question is: "How and where can we produce the necessary energy by renewable resources?"

In the scenarios mentioned above, about 2,000 GW solar, 650 GW wind, together with hydropower and other renewable energy resources, could produce 5,000 TWh electricity in 2050. Green hydrogen needs to be produced by additional green electricity production capacity in Europe over and beyond the 2,000 GW solar and 650 GW wind capacity. Far offshore wind in the North Sea, Baltic Sea, Mediterranean Sea and in the Atlantic Ocean can produce cost-competitive green hydrogen by transporting this hydrogen to the shore by pipeline. Next to this, wind combined with solar on good locations in Southern Europe (Spain, Italy, Portugal and Greece) could produce cost competitive hydrogen too.

However, a substantial part of the necessary hydrogen needs to be imported from neighboring regions. We currently import a substantial part of our energy from Russia, but North Africa, where green hydrogen can be produced even at lower cost than in Europe and transported through cost-effective pipelines, requires due consideration. In North Africa the solar resources are even better than in Southern Europe and several areas have world class wind speeds. Many countries have ample space available to produce green electricity and hydrogen for their own consumption, but certainly also to export to Europe and even beyond.

Green hydrogen can be imported by ship as liquid hydrogen, ammonia (NH_3) or methylcyclohexane (MCH, hydrogen bound to toluene), from additional sources further away, like LNG nowadays (IEA, 2019). But at distances below 4,000 km, shipping is more expensive than pipeline transportation (Lanphen, 2019). Therefore, we only consider hydrogen pipeline transport between Europe and North-Africa.

Energy supply and demand need to be balanced at all time. The large seasonal fluctuations in energy demand, especially for space heating, and the variability of solar and wind, require large scale storage capacity. We assume that 20% of the final energy demand, both for electricity and hydrogen, needs to be supplied via hydrogen energy storage. 1200 TWh final electricity demand must therefore be supplied by electricity production from stored hydrogen. This electricity can be produced by fuel cells, placed close to the demand centers. Expensive electricity transport cost will be avoided and the excess heat from these fuel cells could be used for space heating, if feasible.

An Energy Balance can be constructed, considering the necessary storage volumes (3,630 TWh or 92 million ton hydrogen), the conversion losses from electricity to hydrogen (20%), hydrogen to electricity (40%), the transport losses for electricity (5%) and hydrogen (2%) and the hydrogen storage losses (10%), see Fig. 10. The necessary primary energy (electricity) to deliver 12,000 TWh final energy demand, is 15,710 TWh. This yields an overall system efficiency of 76%, a little better than today's 72%. However, in a fully renewable energy system, not based on finite energy sources, this efficiency figure loses relevance. In the future, it is not about system efficiency, but about system cost.

Important to note is that from this 15,710 TWh primary energy production, 8,450 TWh, 54%, is produced in Europe and 7,260 TWh, or 46%, is produced in North Africa. It should be noted that this is a marked improvement over our current situation, where Europe imports 55% of its primary energy demand. Considering the annual hydrogen production, 8,520 TWh in total, more than 2/3 needs to be imported from North Africa, see Table 6. It clearly shows that for a fully renewable energy system in

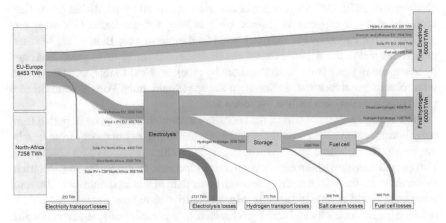

Fig. 10 Energy balance European Union 2050; primary energy production is 15,710 TWh, with final energy consumption amounting to 12,000 TWh, 50% electricity and 50% hydrogen

Table 6 Primary energy production in Europe and North Africa, for use in the EU in 2050

2050 Primary energy production	Electricity capacity (GW)	Electricity production (TWh)	Hydrogen production (TWh)	Hydrogen production (Mton H2)
Europe				
Solar PV	2,000	2,800		
Wind (onshore + offshore)	650	2,010		
Hydro + other renewables		240		
Additional offshore wind for hydrogen	600	3,000	2,400	61
Additional Wind + Solar PV for hydrogen	100	400	320	8
Total Europe		8,450	2,720	69
North Africa				
Solar PV	2,000	4,400	3,520	89
Wind (onshore)	500	2,000	1,600	41
Solar PV + CSP (hybrids)	170	860	690	17
Total North Africa		7,260	5,810	147
Total		15,710	8,530	216

Europe, we need North Africa to produce cost-competitive solar and wind electricity, converted to hydrogen, for export by pipeline to Europe.

What Needs to be Done?

Such a renewable energy scenario for Europe, implies a massive program to realize renewable energy capacity both in Europe and in North Africa. About 4,200 GW solar capacity and 1,800 wind capacity needs to be realized. To appreciate the enormous investment, the current installed capacity of all coal fired power plants in the world amounts to 2,000 GW. For conversion from electricity to hydrogen, about 3,400 GW of electrolyzer capacity is needed. And for the conversion from hydrogen to electricity, about 500 GW of fuel cell capacity needs to be installed, see Table 7.

Next to this, the pipeline capacity between North Africa and Europe needs to be expanded to about 1,000 GW, which is about 30 pipelines with a capacity of 33 GW each. In Europe the gas pipeline infrastructure is partly available. However, the production flows of hydrogen are, to a large extent, from the south to the north. This means that, especially from the south of Europe to the north of Europe, the

Table 7 Capacities in production, conversion, infrastructure and storage that needs to be realized in 2050

2050	To be realized in 2050 (to cater for Europe's energy demand)
Production	
Solar capacity	4,200 GW; 2,000 GW Europe and 2,200 GW North Africa
Wind capacity	1,800 GW; 1,300 GW Europe and 500 GW North Africa
Conversion	
Electrolyzer capacity	3,400 GW; 700 GW Europe and 2,700 GW North Africa
Fuel cell capacity	500 GW; in Europe
Infrastructure	
Hydrogen pipelines	1,000 GW pipeline connection between Europe and North Africa Re-use existing gas pipeline infrastructure in Europe and North Africa, conversion from natural gas to hydrogen Expand pipeline capacity, especially from south to north of Europe Realize pipeline connections between North African countries from east to west
Electricity grid	Massive capacity expansion of the electricity grid, at least with a factor 2 on a volume base Grid re-enforcement and new grids are required between the renewable electricity production in Northern and Southern Europe and the load centers Capacity expansion of interconnections between countries Realize an electricity grid in and between North African countries
Storage	
Salt caverns	15,000 Salt caverns; 10,000 in Europe, 5,000 in North Africa Hydrogen storage in empty gas fields if possible
Batteries	Batteries are required, especially for day-night storage; North Africa could rely much more on battery storage than on hydrogen storage, due to its climatological conditions
Heat storage	Seasonal heat and cold storage for space heating, especially in aquifers and rock formations. This is important for North-Europe

hydrogen gas grid needs to be expanded. In Europe also the electricity transport infrastructure needs to be considerably expanded. Final electricity consumption in 2017 was about 3,000 TWh and will be doubled in 2050. On a volume basis, the capacity therefore needs to be expanded with at least a factor of 2. However, especially the grid capacity from the north and south of Europe to the load centers and the interconnections between the European countries need be expanded much more than a factor of 2, see Table 7.

The estimated large-scale storage volume in this scenario is 3,630 TWh or 92 million ton hydrogen. One salt cavern can store about 6,000 ton of hydrogen. So, there is a need for 15,000 salt caverns for hydrogen storage. Although these salt caverns can be realized in Europe, also in North Africa salt cavern storage capacity needs to be realized. 1/3 of the salt cavern capacity needs to be realized in North

Africa. Next to this there is a need for battery storage, especially for day-night storage and heat and cold storage, especially seasonal storage in aquifers, see Table 7.

An important aspect of this transition is that the end-use conversion technologies need to be replaced. It means e.g., the replacement of all present internal combustion engine vehicles by electric vehicles, both battery and fuel cell hydrogen electric vehicles. But also replacing existing heating equipment in houses, with heat pumps, hydrogen boilers or fuel cells. However, this equipment is replaced every 10–15 years anyhow. Therefore in 30 years' time the replacement of this equipment is manageable.

Towards a Sustainable and Inclusive Energy System

An inclusive energy transition is about an energy system that is affordable, accessible, secure, reliable and fair (distribution of benefits and burdens) for everyone. A 50% renewable electricity and 50% renewable hydrogen system developed in mutual co-operation with North Africa for the benefit of both, whereby everyone is connected to an energy infrastructure including energy storage facilities (electricity, hydrogen), is a good prerequisite for an inclusive energy system.

Affordable

Renewable electricity is rapidly becoming cheaper than conventional electricity made in nuclear, gas- or coal-fired power plants. Already to date, solar power in Southern Europe and offshore wind in the North Sea does not require subsidy but can be produced and sold at market prices. If a green hydrogen market would develop along the lines sketched here, hydrogen can be produced at € 1 per kg, which is compatible with natural gas prices of € 9/mmbtu or € 0.25/m^3. Since the energy content of 1 kg of hydrogen is equivalent to 3.8 L of gasoline, it is also cheaper than gasoline or diesel at that price point, even discounting the tax on transport fuels.

The advantage of a mutual co-operation with North Africa are two-fold: the economic opportunity for North African economies, and the lower production cost for hydrogen from solar and wind electricity. The resources are better, investment costs lower and space is abundantly available. Hydrogen could be produced for less than € 1 per kg and be competitive with hydrogen produced in Europe, even including pipeline transportation cost.

But the main advantage lies in the infrastructure and storage. The proposed transition would, to a large extent, use the existing natural gas grid and would avoid an expensive and troublesome complete overhaul and large capacity expansion of the electricity grid. Also, storage, especially large-scale seasonal storage for hydrogen can be realized similar to natural gas, in existing and newly realized salt caverns. Hydrogen storage in salt caverns can not only be realized much cheaper, it can be realized faster too.

An affordable energy system for everyone is not necessarily a system where energy is produced, stored and consumed locally. Especially renewable energy resources show a great variation in production cost around the world. At places with good solar irradiation or wind speeds and cheap land and competitive labour cost, the production cost could be a factor 5–10 lower than at places with moderate solar irradiation or wind speeds, and high land and labour costs. Also, large scale energy storage costs (hydrogen in salt caverns or ammonia in large tanks) are easily a factor of 100 cheaper than small scale storage costs (compressed hydrogen in bottles or electricity in batteries). Therefore, an affordable energy system for everyone will be a smart combination of a large scale and local energy system.

Accessible

An accessible energy system for everyone is a system whereby everyone has access to clean energy. An important pre-requisite for access to clean energy is a connection to a well-organized energy infrastructure. This could be an electricity, gas or heat grid or a fuelling infrastructure, whereby an electricity grid connection is the most essential one. A combination with a gas infrastructure (hydrogen), especially in Europe, is in many cases useful to deliver the necessary energy for heating at moments of high demand.

The connection to an energy infrastructure needs to be guaranteed. This could be organized by obliging energy transport and distribution companies to connect every consumer and every producer. The question is whether the cost for such a connection needs to be socialized. This seems to be a fair principle, making it possible that all consumers are not only connected to an energy infrastructure, but also could afford to pay for it.

The obligation to connect and socializing energy infrastructure cost seems a good principle, but how to implement these principles in a fully renewable energy system, with electricity and hydrogen as the main energy carriers? Two types of questions arise: connecting to which infrastructure and socializing over what energy? We illustrate these questions with two examples, informing the debate for policy makers.

- If a far offshore wind farm is realized, is there an obligation to connect to an electricity grid or can the wind farm owner choose to connect to either the electricity or the hydrogen grid? Or can the energy transport companies (TSO's) together decide to connect to a hydrogen grid and/or an electricity grid?
- If the natural gas infrastructure is converted to a hydrogen infrastructure, do we socialize the cost over the hydrogen consumption only, or do we socialize these costs in a transition period over the total gas (natural gas and hydrogen) consumption? Or should we socialize the cost for all infrastructure (electricity, natural gas, hydrogen) over all energy consumption?

Secure

Security of supply is always an important consideration for Europe, especially because energy is a vital part of the economy. Europe is currently a net energy importer and will likely continue relying on imported energy for a share of its demand, also in a future renewable energy system. However, a system as described here substantially reduces the import share of currently 55% to 46%, with a more diverse set of countries supplying Europe.

The infrastructure proposed also carries important benefits for North African nations, enabling them to secure their own energy supply and trade hydrogen and electricity among each other and exporting to Europe and other parts of the world, earning foreign exchange and boosting their economies.

Reliable

To deliver energy at the right time and place, an energy infrastructure with enough transport and distribution capacity is necessary together with enough storage capacity at different time scales (seasonal, weekly, daily, hourly, minutes and seconds). An all-electric system, whereby only electricity is transported, stored and distributed, needs a gigantic and very expensive expansion of the electricity grid and battery storage capacity. In Europe especially, with seasonal storage needs due to space heating, such an all-electric system seems prohibitively expensive and almost impossible to realize. In North-Africa, however, with less seasonal variation and where solar production matches cooling demand, an all-electric solution with battery storage to cater for the evening peak, seems a good and cost-effective solution. At the same time, demand for hydrogen for mobility and industry will also develop in North Africa.

In Europe, due to its existing natural gas infrastructure, a smart combination of a green electricity and hydrogen energy system could offer a reliable and cost-effective solution. Of course, there is a need to modernize and increase the capacity in the electricity grid, together with installing battery and capacitor capacity for frequency response and short-term storage. But especially for weekly and seasonal storage, hydrogen offers a much cheaper solution, especially by storing hydrogen in salt caverns.

Fair

The development of a clean energy system for both Europe and North Africa in mutual co-operation is beneficial for both. Europe cannot produce the renewable energy it needs in Europe alone, as it simply does not have enough solar and wind resources, nor available and affordable land. North Africa on the other hand, has these resources abundantly available and can produce enough clean energy for its own demand, as well as for export to Europe and other parts of the world. North Africa, however, lacks the technology, capital and a well-educated labour force to develop a clean energy system on its own.

Therefore, cooperation on the development of a renewable, fully decarbonized energy system is for the benefit of both. It creates future-oriented economic development, jobs and welfare in North Africa by developing a clean energy system for their own use and export. In Europe, a clean, reliable and affordable energy system can be realized by re-using part of the existing assets and infrastructure, in combination with renewable energy production in Europe and import from North Africa. Europe can build a sustainable, circular and cost-competitive industry, based on green electricity and green hydrogen supply at competitive cost. And Europe and North Africa together could be world market leaders in renewable energy system technology and system production and realization, especially in electrolyzer, gasification and fuel cell technologies, hydrogen, electricity and heat storage technologies, energy infrastructure and conversion technologies, green chemistry and synthetic fuels.

Improving livelihoods in Africa will reduce the migration of people from Africa, seeking economic opportunities in Europe. The joint development of a clean energy system could provide a perspective for a better life and future in these North African countries. Such a development is fair from both the European and North African perspective.

Conclusions and Required Political Agenda

A European energy system based on 50% green electricity and 50% green hydrogen, developed in mutual co-operation with North Africa for the benefit of both, would have many advantages:

- The energy system would be entirely clean, with no CO_2 emissions, which meets the Paris Agreement but would also have tremendous health benefits due to reduced local emissions in European and African cities.
- The system would be a shift away from a system based on finite resources, which invariably leads to scarcity and higher cost towards the end, to a system entirely based on renewable energy resources with technologies becoming cheaper over time.
- A European ambition level based on proven but largely undeveloped technologies (electrolyzers, gasifiers, fuel cells, hydrogen storage technologies, new domestic appliances, hydrogen-electric mobility, synthetic green chemicals and fuels) provides a tremendous opportunity for global technology leadership, with associated economic momentum and job creation.
- The infrastructure required for the new system will be largely based on the already existing natural gas grid and avoids an expensive overhaul and massive capacity expansion of the electricity grid.

- The energy system would be reliable, with balanced supply and demand at all times and every place, due to large-scale, cheap hydrogen storage, especially in salt caverns, together with a public hybrid electricity-hydrogen infrastructure.
- Developing a clean energy system in cooperation between Europe and North Africa unlocks access to vast and cheap renewable energy resources for Europe and North Africa, whilst supporting the development of affordable, reliable and clean energy for North Africa itself.
- Europe and North Africa can both profit from this cooperation, it creates economic development, new business, new export, jobs and welfare in North Africa as well as in Europe
- Developing a clean energy system in North Africa, for own use and export, creates jobs, welfare and better living conditions reducing the necessity for people to migrate to Europe.

However, such a "moonshot" program requires tremendous political and societal will on a level rarely seen, not only within Europe but also between Europe and North Africa. To enable the transition and avoid the exclusion of large parts of the current energy industry, careful thought must be given to minimize stranded assets and include as many players as possible. An environment for investments in Europe and North Africa needs to be designed, in mutual co-operation, for the benefit of both.

The following are necessary considerations for an action agenda:

- A strong, clear and lasting political commitment is necessary, embedded in a binding European strategy with clear goals stretching over several decades.
- A new type of public private partnership on a pan-European level must be crafted, with the aim to create an ecosystem to nurture a European clean energy industry that has the potential to be world leading in the field. This partnership should include existing energy industry as well as innovative newcomers.
- A novel enabling regulatory environment and associated market design is required for the necessary investments, whilst keeping the system costs affordable.
- An integrated electricity-hydrogen infrastructure and storage system policy framework needs to be designed, with fair and reliable access to energy for everyone.
- Finally, above all, a new, unique and long-lasting mutual cooperation on the political, social and economic level between the EU and North Africa needs to be designed and realized. This cooperation needs to be based on mutual respect and trust, considering each other's cultural, social and economic backgrounds.

References

BP. (2019). "BP energy outlook, 2019 edition."

Bundesnetzagentur. (2019). "Quartalsbericht zu Netz-und systemsicherheitsmaßnahmen; Gesamtjahr und viertes quartal 2018." May. https://www.bundesnetzagentur.de/SharedDocs/Downloads/DE/Allgemeines/Bundesnetzagentur/Publikationen/Berichte/2019/Quartalsbericht_Q4_2018.pdf?__blob=publicationFile&v=4

Bünger, U., Michalski, J., Crotogino, F., & Kruck, O. (2016). Compendium of hydrogen energy. In *Large scale underground storage of hydrogen for the grid integration of renewable energy and other applications*, (pp. 133–163). Woodhead Publishing.

Caglayan, D., Weber, N., Heinrichs, H. U., Linßen, J., Robinius, M., Kukla, P. A., & Stolten, D. (2020). Technical potential of salt caverns for hydrogen storage in Europe. *International Journal of Hydrogen Energy, 45*(11), 6793–6805. https://doi.org/10.1016/j.ijhydene.2019.12.161

Carbon Capture and Storage, Association. (2019). https://www.ccsassociation.org/faqs/ccs-glo bally/

Dii, & FraunhoferISI. (2012). *"Desert power 2050 perspectives on a sustainable power system for EUMENA."* Presented at SWP Berlin, 22 May.

DNVGL. (2017). "Verkenning waterstofinfrastructuur (in Dutch)." Rapport for Ministery of Economic Affairs.

DNV-GL. (2018). "Energy transition outlook 2018, a global and regional forecast of the energy transition to 2050." https://eto.dnvgl.com

European Commission. (2018). "A clean planet for all; A European strategic long-term vision for a prosperous, modern, competitive and climate neutral economy." Brussels.

Eurostat. (2019). 7 February. https://ec.europa.eu/eurostat/documents/2995521/9549144/8-070 22019-AP-EN.pdf/4a5fe0b1-c20f-46f0-8184-e82b694ad492

EurostatEnergy. (2019). "EU28 energy balances february 2019 edition." https://ec.europa.eu/eur ostat

Eurostatgas. (2019). "Natural gas supply statistics; Consumption trends." https://ec.europa.eu/eur ostat/statistics-explained/index.php?title=Natural_gas_supply_statistics#Consumption_trends

EurostatGuide. (2019). "Energy balance guide draft." 31 Januari. https://ec.europa.eu/eurostat/ documents/38154/4956218/ENERGY-BALANCE-GUIDE-DRAFT-31JANUARY2019.pdf/cf1 21393-919f-4b84-9059-cdf0f69ec045

Eurostatimports. (2019). "EU import of energy products—recent developments." May. https://ec. europa.eu/eurostat/statistics-explained/index.php?title=EU_imports_of_energy_products_-_rec ent_developments

EWEA. (2017). "2050: Facilitating 50% wind energy." https://www.ewea.org/fileadmin/files/lib rary/publications/position-papers/EWEA_2050_50_wind_energy.pdf

FCHJU. (2019). "Hydrogen roadmap Europe, a sustainable pathway for the European energy transition." https://fch.europa.eu

Hydrogen Council. (2017). "Hydrogen scaling up; a sustainable pathway to the global energy transition." November. https://hydrogencouncil.com/study-hydrogen-scaling-up/

HyUnder. (2013). "Assessment of the potential, the actors and relevant business cases for large scale and seasonal storage of renewable electricity by hydrogen underground storage in Europe." EU report 14 August 2013 www.fch.europa.eu

IEA. (2019). "The future of hydrogen, seizing today's opportunities." Report prepared by the IEA for the G20 Japan.

IRENA. (2015). "Africa 2030; roadmap for a renewable energy future." https://www.irena.org/pub lications/2015/Oct/Africa-2030-Roadmap-for-a-Renewable-Energy-Future

James, B., DeSantis, D., Huya-Kouadio, J., Houchins, C., & Saur, G. (2018). "Analysis of advanced H2 production & delivery Pathways." June. https://www.hydrogen.energy.gov/pdfs/review18/ pd102_james_2018_p.pdf

Kiwa. (2018). "Toekomstbestendige gasdistributienetten (in Dutch)." Rapport for netbeheer Nederland.

Lanphen, S. (2019). "Hydrogen import terminal; providing insights in the cost of supply chain of various hydrogen carriers for the import of hydrogen." MsC thesis, TU Delft, Delft.

Michalski, J., Büngerz, U., Crotogino, F., Donadei, S., Schneider, G. S., Pregger, T., Cao, K. K., & Heide, D. (2017). "Hydrogen generation by electrolysis and storage in salt caverns: Potentials, economics and systems aspects with regard to the German energy transition." *International Journal of Hydrogen.*

Morgan, S. (2019). "Five EU countries call for 100% renewable energy by 2050." 5 March. https://www.euractiv.com/section/climate-strategy-2050/news/five-eu-countries-call-for-100-renewable-energy-by-2050/

Nordstream. (2014). "Secure-energy-for-europe-full-version." https://www.nord-stream.com/media/documents/pdf/en/2014/04/secure-energy-for-europe-full-version.pdf

Nordstream. (2017). *Factsheet the nordstream pipeline project.* https://www.nord-stream.com/

Ram, M., Bogdanov, D., Aghahosseini, A., Gulagi, A. S., Oyewo, A., Child, M., Fell, H.-J., & Breyer, C. (2017). *Global energy system based on 100% renewable energy—Power sector.* Study by Lappeenranta University of Technology and Energy Watch Group.

Shell. (2018). "Shell scenario SKY, meeting the goals of the Paris agreement." https://www.shell.com/skyscenario

Timmerberg, S., & Kaltschmitt, M. (2019). Hydrogen from renewables: Supply from North Africa to central Europe as blend in existing pipelines—Potentials and costs. *Applied Energy, 237,* 795–809

van Wijk, Ad., & Chatzimarkakis, J. (2020). *Green hydrogen for a European green deal, a 2x40 GW initiative.* Hydrogen Europe. ISBN 978-90-827637-1-3

van Wijk, A., van der Roest, E., & Boere, J. (2017). *Solar power to the people.* Allied Waters. ISBN 978-1-61499-832-7 [online]. https://doi.org/10.3233/978-1-61499-832-7-i

Varadi, P. F., Wouters, F., & Hoffmann, A. R. (2018). *The sun is rising in africa and the middle east—On the road to a solar energy future.* Pan Stanford Publishing Co ISBN-10-9814774898.

Decentralised Control and Peer-To-Peer Cooperation in Smart Energy Systems

Geert Deconinck

Abstract In order to achieve a decarbonised energy system, change has to happen from electricity generation to the transmission grid over the distribution level all the way down to the industrial loads and the local households. To get involvement of communities in this energy transition, local participation is needed, so that the citizens can be aware of the impact of their energy-related actions on environment and climate. However, the energy system has typically been organised in a top-down fashion, with centralised approaches and little active control, resulting in passive grid and ditto customers. Smart grids have put active customers and consumer engagement as one of the cornerstones of a more intelligent energy infrastructure, which can be organised differently. Indeed, in different niches decentralised approaches have been used successfully (decoupled microgrids, peer-to-peer networks, etc.). This chapter explores how decentralised approaches can fit the future energy system and how it can empower people for engaging in the energy transition.

The Rise of Local Energy Communities

Decentralisation is becoming an important paradigm within energy systems. This happens not only at the physical level, where electricity generation decentralises by the widespread introduction of photovoltaic installations or other distributed energy resources or where heat is produced locally, close to where it is used. Also, data infrastructures become omnipresent with embedded monitoring and control systems, allowing for local smart energy applications (such as microgrids, virtual power plants or home energy management systems). Besides the technical aspects also policy opens up to local initiatives, in order to get the energy user more engaged in the energy transition and to get radical innovations (Lavrijssen & Carrillo Parra, 2017).

This decentralisation paves to way towards local energy communities, and—in its most distributed version—to peer-to-peer energy trading. In a local energy community, a group of households or consumers shares some energy assets (like a battery

G. Deconinck (✉)
KU Leuven/EnergyVille, Leuven, Belgium
e-mail: geert.deconinck@kuleuven.be

© The Author(s) 2021
M. P. C. Weijnen et al. (eds.), *Shaping an Inclusive Energy Transition*,
https://doi.org/10.1007/978-3-030-74586-8_6

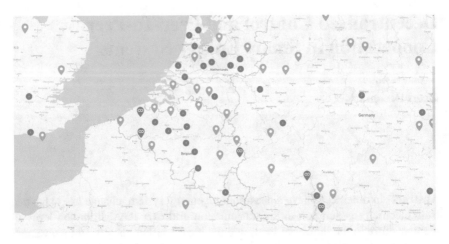

Fig. 1 Energy cooperatives in Belgium and in parts of the Netherlands, France, UK and Germany, that are member of REScoop (REScoop.eu (2020))

storage system or a heat buffer, or some electricity generation devices) and they organise themselves independently from supplier or distribution system operator. In a peer-to-peer energy system, customers buy and sell energy directly among each other. Such energy communities are often organised in energy cooperatives as legal entities.

Approximately 1,500 of these European energy cooperatives, representing about 1,000,000 citizens are gathered in the Renewable Energy Cooperatives (REScoop) (REScoop.eu, 2020) (Fig. 1). It REScoop describes energy communities as "a way to organise citizens that want to cooperate together in an energy-sector related activity based on open and democratic participation and governance, so that the activity can provide services or other benefits to the members or the local community. […] The primary purpose of energy communities is to create social innovation - to engage in an economic activity with non-commercial aims." (REScoop.EU, 2019). As an example, some of the cooperatives invest in photovoltaic or wind generation for their participants, or organise joint buying of insulation material to improve energy efficiency, others cooperate with social housing companies to provide energy services or to renovate the building stock.

Although some forms of fully decentralised control (such as directly selling electricity to your neighbour over the public domain) are not (yet) allowed from a regulatory perspective in many countries, it is clearly the way forward. Such decentralisation has also been proposed in 2018 by the European Commission in its 'Clean Energy for all Europeans' package, which defines 'Citizen Energy Communities' as one of the cornerstones in the energy transition.[1] Citizen Energy Communities are

---

[1] In that same package, also Renewable Energy Communities are defined in the recast Renewable Energy Directive (EU) 2018/2001, in article 2(16); see (REScoop.EU, 2019) for the subtle differences between these 'citizen energy communities' and 'renewable energy communities'.

explicitly mentioned in article 2(11) of the recast Electricity Directive (EU) 2019/944 (European Parliament, 2019). They are defined as "a legal entity that:

- is based on voluntary and open participation and is effectively controlled by members or shareholders that are natural persons, local authorities, including municipalities, or small enterprises;
- has for its primary purpose to provide environmental, economic or social community benefits to its members or shareholders or to the local areas where it operates rather than to generate financial profits; and
- may engage in generation, including from renewable sources, distribution, supply, consumption, aggregation, energy storage, energy efficiency services or charging services for electric vehicles or provide other energy services to its members or shareholders." (European Parliament, 2019).

Much of the European policy needs to be translated into country-specific implementations, and not all member states are equally front-running in the energy transition. Roberts and Gauthier reviewed all National Energy and Climate Plans (NECP) of the European member states for their view on energy communities and they conclude, as an understatement, that there is much room for improvement (Roberts & Gauthier, 2019). Their key takeaways are that awareness in EU member states on energy communities is moderate but actual planning is very low. At the positive end, a few member states (such as Greece) show strong commitment and their NECP has a comprehensive treatment of energy communities, including defined targets and detailed policies and measures. At the negative end, other member states (such as Estonia, Germany, Malta and Sweden) completely ignore energy communities in their NCEP. In between, many NECPs suffer from ambiguity around energy communities and they fail to distinguish them from distinct activities such as (individual) self-consumption. They conclude by stating that in the NECPs, "renewable energy communities and self-consumption overshadow other dimensions where energy communities can contribute: energy efficiency, energy poverty, ownership of distribution network, e-mobility, rural development, district heating, etc." (Roberts & Gauthier, 2019).

Often the energy cooperatives or local energy communities are actively looking for a role in the traditional energy landscape, where distribution system operators as regulated entities often take up a rather conservative role. In such context, these innovative communities have to find their place between existing players (such as suppliers or retailers) and new players (energy service providers, aggregators, flexibility service providers, etc.).

All of these evolutions push also the *control* in the smart energy system towards lower and more decentralised levels.

Bottom-Up Control and Top-Down Control Meeting in the Middle

Traditionally the power system has been controlled in a centralised way, with a prominent role for the system operator (e.g., the transmission and distribution system operator in a European context, or the independent system operator in a North-American context). Their role is to organise markets, and make sure that the operations are executed fluently by activating the necessary reserve power plants or decreasing power production to keep the grid balanced (ENTSOE, 2019). Keeping the frequency and voltage at their nominal levels and avoiding congestions are important assignments for this top-down control paradigm. While frequency is mainly a transmission grid problem, voltage and power quality issues appear everywhere, also at distribution level. However, at the distribution level (medium and low voltage level), there are not many controllable actuators, and the grid is often operated in quite a passive way. The distribution system operator has control over a number of devices in the substations in order to change configurations (connecting feeders and loads to particular bus bars), to change voltage levels (by changing the transformation ratio of transformers), or to change protection settings (to change selectivity in the protection). However, there is barely any control possible beyond the substations in the feeders. Even the *monitoring* at that level is limited as there are little or no sensors (fault indicators, voltage sensors, power quality measurements) in the feeders and sub-feeders. If a problem occurs, e.g., leading to a local blackout, the distribution system operator often has to wait until consumers call the service centre, and crew has to be sent in the field in order to solve the problems.

The trend towards smart grids implies that also the low and medium voltage grid gets equipped with sensors for monitoring and actuators for control. Because of this, faults can be detected earlier and some problems can be solved remotely without having to send crew in the field, so that the downtime for customers can decreased. Still this control is very much top-down and centralised in the operation room of the distribution system operator.

With more renewables at the customer's premises, also control has been added to that local level: the inverters that connect e.g., photovoltaics to the home, have a *local* embedded controller that continuously measures the grid voltage and disconnects the inverter in case there is an overvoltage. The national or regional grid codes specify (e.g., (Synergrid, 2017)) at which voltage level small photovoltaic systems (less than 10 kVA) need to disconnect. For larger installations, the ratio between active and reactive power supplied by the inverter can be used to support the voltage in the grid to a larger extent, and also here distribution system operators prescribe particular grid support requirements.

However, when active customers are envisioned, there is much more need (and opportunity) for monitoring and control at the local level, even behind the point-of-common-coupling, i.e., behind the meter that connects an individual customer to the grid. This is a fortiori the case when a smart meter is installed, which digitally measures off-take from the grid and injection in the grid, and has the possibility to communicate this information to a data operator, or to the distribution system operator, from where third parties (e.g., suppliers or retailers) have access to it. Smart meters allow for better monitoring or, if connected to actuators, for control applications as well. First, the smart meter allows to measure local power quality issues, or allows obtaining a local load profile from which the customer can benefit when negotiating a contract with his supplier. Additionally, local flexibility (at a residential level or at an industrial customer connected to the low or medium voltage grid) can be remotely controlled by an aggregator, which valorises it on the energy market or on the ancillary services market. Such flexibility results from the ability to shift the electricity use over time, and this increase or decrease in local off-take has a value at an aggregated level (Vandael et al., 2013). Another example is provided by a battery owner, who can use the local controller that manages the charging and discharging of the battery, to maximise its self-consumption, or to use the battery flexibility for an aggregator. In general, these *demand response* applications provide a great opportunity to better match the demand of electricity to its supply (Deconinck & Thoelen, 2019). A good survey of the state-of-art on demand response applications can be found in (Shoreh et al., 2016; Siano, 2014).

Hence, in a future smart grid, one sees many different control architectures appearing *in parallel*. Some of them are top-down, as for the distribution system operator controlling reactive power settings of renewable energy sources, or the data operator reading out smart meter registers, or the aggregator controlling flexibility. Others are local controllers that control the battery or interact with a home energy management system. These are examples of bottom-up control systems that generate a local equilibrium state that need to be taken as granted (or as un-controllable) by the top-down control applications.

It is foreseen that the amount of such control structures will only increase in the future (as e.g., new regulations will allow to have a contract with different suppliers simultaneously—one for injection in the grid and one for grid offtake (Atrias, 2019). Having all of these control architectures in parallel (especially if they do not share a common communication infrastructure) does not provide an optimal use of resources. Given the enormous ecological footprint of information and communication technology, and the need for digital sobriety (Ferreboeuf, 2019), it would be better if some of these infrastructures could be shared.

Such shared communication and control infrastructure rely on *interoperability*. Interoperability can be defined as the ability of two or more devices from the same vendor, or different vendors, to exchange information and to use that information for correct cooperation (International Electrotechnical Commission, 2019). Such interoperability needs to be effective at different levels: at the level of the communication media and protocols, at the level of the information model, at the functionality level and at the business level. The Smart Grid Reference Architecture Model (Fig. 2) provides an adequate way to reason about such interoperability for smart grid applications, in a layered approach (Smart Grid Coordination Group CEN-CENELEC-ETSI, 2012).

Instead of having all these parallel control architectures, it would be beneficial if entities (smart meters, home energy management systems, batteries …) would be

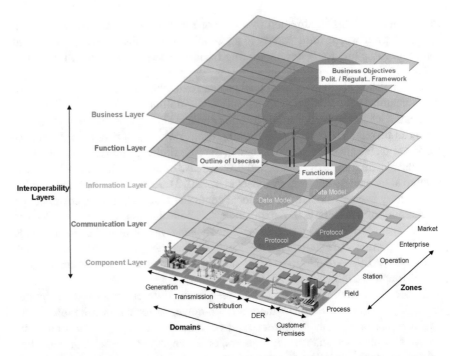

Fig. 2 SGAM framework (Smart Grid Coordination Group CEN-CENELEC-ETSI, 2012)

able to participate in different smart grid applications at the same time, because of an interoperable interface.

Additionally, from the viewpoint of the privacy of the end user, and of the confidentiality of its data, it would be better if local data is kept and processed locally, and is not transferred along the communication infrastructure several times for different applications. Local data provides better privacy protection than central data, and a single source for storing the data is less error prone than multiple sources for the same pieces data.

This interoperability would enable a more active engagement of the customer, including the ability to work with multiple market players at the same time (multiple suppliers, multiple aggregators, within a local energy community, etc.,). It hence enables end-customers to participate in multiple smart grid applications according to their needs, in an open market context.

This brings us to the question, which is the most appropriate distribution paradigm for the control actions, for which we introduce the taxonomy of Fig. 3. In a case with only *local* control, there is no communication between devices, and control decisions are taken only based on locally measured parameters (voltage, power, known price profiles, etc. Often it is beneficial however to *coordinate* control between entities, such that it is not only based on local sensor information, but also on information from other entities that is communicated to them. This can be done in different ways.

Fig. 3 Relevant control
paradigms for local and
coordinated control
(Deconinck & Thoelen,
2019)

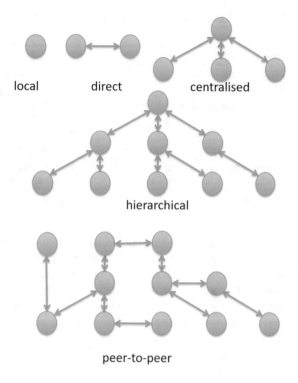

- In *direct* control there is a direct interaction between two entities, for instance an intelligent electric vehicle charger that determines the optimal moment to charge the car.
- With *centralized* control, a central agent controls all flexibility, for instance a distribution system operator that sends an interruption signal to specific loads to be shed.
- With a *hierarchical* approach, there are intermediate levels that ensure some scalability.
- Within a *peer-to-peer* approach, components only interact with some physical or logical neighbours in a flat hierarchy.

Table 1 indicates relevant differences and advantages of the different approaches.

Going Completely Decentral: Peer-To-Peer

Opposite to the centralised control, completely decentralised systems pop up at the other end of the spectrum: there, each unit is autonomous and collaborates with its neighbours, in a peer-to-peer fashion. More than a decade ago, it has been shown—both in simulations and in a practical implementation—that such completely

Table 1 Advantages of centralised and decentralised control

Centralised control	Decentralised control
More simple	More scalable
Single point of control	No single point of failure
SCADA-compatible Requires dedicated communication architectures Master/slave protocols	Internet-compatible Fits with many communication architectures: overlay networks, peer-to-peer ... Publish/subscribe protocols
One control structure per application	Each entity interfaces to multiple applications
More compatible with integrated energy companies with few actors	More compatible to a liberalized, open market model with many actors
Passive customers	Engaged, active customers

distributed approaches are a feasible alternative to control a microgrid with distributed energy resources as an *autonomous electricity network (AEN)* without any central master (Vanthournout et al., 2005). In that microgrid, control was implemented as a combination of *local* droop control at the inverters, with a decentralised secondary control (to keep the voltage within its limits), and a decentralised tertiary control application (that prioritises the more economic energy resources) (Brabandere et al., 2007). The communication infrastructure was based on a peer-to-peer gossiping protocol that ran on top of a dynamic semantic overlay network (Deconinck & Vanthournout, 2009).

Many more examples of decentralised or peer-to-peer control in smart grid applications do exist meanwhile. In (Almasalma et al., 2017), a distributed voltage control and optimisation method is proposed that optimises the reactive power settings of selected resources in a microgrid to keep the voltage quality within predefined limits. Also here, communication is fit for a decentralised approach, with a gossiping protocol to spread information between the active component, and with a distributed optimisation scheme based on dual decomposition to calculate the voltage set points. A hardware-in-the-loop implementation of this approach validated its performance, using a wireless device-to-device communication system (Almasalma, 2018). Its setup is represented in Figs. 4 and 5. This implementation showed that fully distributed voltage control systems can indeed provide satisfactory regulation of the voltage profiles.

A larger case study on a 62-bus and 124-bus network provided very acceptable results in terms of both convergence speed and optimality (Almasalma et al., 2019).

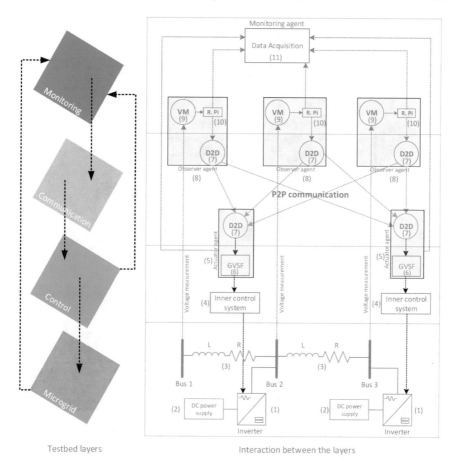

Testbed layers   Interaction between the layers

Fig. 4 Multi-layer multi-agent architecture of the peer-to-peer (P2P) voltage control testbed (VM stands for voltmeter, D2D: device-to-device communication module, GVSF: grid voltage support function, R.Pi: raspberry pi computer, R: resistor, L: inductor, labels 1 to 11 indicate the different parts of the testbed) (Almasalma, 2018)

Peer-To-Peer Control, Communication and Trading

Not only the control architecture and communication architecture can be decentralised, also business applications can be run in a similar way. In the European H2020 project P2P-SmarTest ("P2P-SmarTest Project", 2019), energy trading between microgrids was seen as major example to develop a decentralised approach on the three levels: control, communication and trading (Pouttu et al., 2017). This energy trading is presented in Fig. 6. Also other researchers have investigated similar approaches of control and energy trading between autonomous entities, whether they call them microgrids, cells (multi-microgrids) or multi-cells (Zhang et al., , 2017, 2018). All of these levels provide different degrees of aggregation.

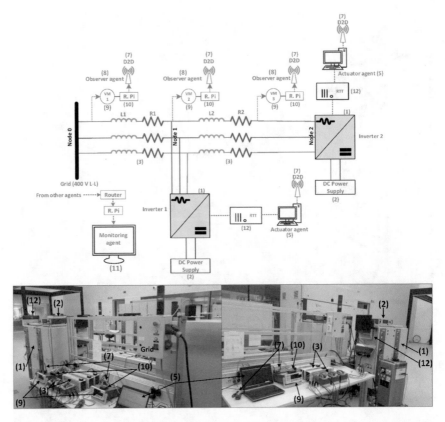

Fig. 5 Schematic of the peer-to-peer voltage control testbed (VM stands for voltmeter, D2D: device-to-device communication module, R: resistor, L: inductor, RTT: real time target computer, R.Pi: raspberry pi computer, labels (1)–(12) indicate the different parts of the testbed, labels (1)–(11) are the same as in Fig. 4) (Almasalma, 2018)

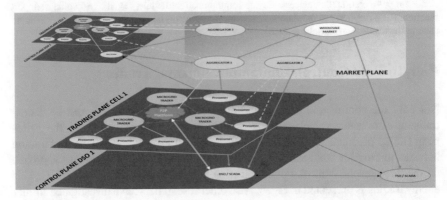

Fig. 6 Peer-to-peer energy trading in the P2P-Smartest project (Pouttu et al., 2017)

Decentralised approaches also allow for different trust models and different ways to validate actions and transactions. In this context, distributed ledger technology that provides a consensus algorithm on top of peer-to-peer networks is an interesting alternative to centralised databases (Deconinck & Vankrunkelsven, 2020). Well-known examples of this technology include blockchains (such as *Ethereum*), block directed acyclic graphs (such as *RChain*) and transaction-based directed acyclic graphs (such as *Byteball*), on top of which *smart contracts* can be implemented. Such smart contracts allow the performance of credible transactions without third parties, in a way that is irreversible, irrefutable and trackable. In the energy sector, different applications of blockchain and smart contracts are being developed (Mihaylov et al., 2018; Peter et al., 2019). In this context, the Users TCP (Technology Collaboration Programme) of the International Energy Agency (IEA) has set up a "Global Observatory on Peer-to-Peer, Community Self-Consumption and Transactive Energy Models" as "a forum for international collaboration to understand the policy, regulatory, social and technological conditions necessary to support the wider deployment of peer-to-peer, community self-consumption and transactive energy models" (GOP2P, 2021). It collects best practices and lessons learned of such decentralised models, beyond the technical perspective, from all over the world.

Having all these trends towards decentralisation, an interesting rhetoric question pops up, whether in a future smart grid, where the local resources in the microgrid are providing grid or where a microgrid manager takes over this responsibility of a central controller, it still makes sense to have a *distribution system operator* to run the grid? Alternatively, the grid codes and the active control units in the different devices are able to provide the same services. In our opinion, this thought experiment needs to be elaborated, in order to see the real minimal control requirements to be allocated to a central unit in a context of interoperable devices and peer-to-peer control applications. Especially if the distribution system operators are not eager to adapt to the new smart grid ecosystems, their role will be taken over by other players; Europe's Joint Research Centre provides a very good overview of the future-readiness of distribution system operators via its DSO Observatory (Prettico et al., 2019).

Data-Driven Approaches and Machine Learning

Together with the increase in data and data streams in a smart grid, we see a gradual shift in the control methodologies that are used in smart grid applications. Model-based control and optimisation techniques have been used for decades, with very good results, also in an uncertain environment. Especially model predictive control, where a receding horizon is considered to base instantaneous control decisions upon, is a very powerful method. However, such model-based approach requires a model of the physical systems that are controlled. White box and grey box modelling use first principle methods or fit parameters to a reduced order representation of smart grid entities, and control and optimisation is often performed by a single controller. Data-driven approaches, on the contrary, do not require a physical understanding of the behaviour to build a model; rather the model can be learned from analysing the data, with techniques from supervised or unsupervised learning. Many applications of machine learning in demand response applications exist; a good overview is given in (Vázquez-Canteli & Nagy, 2019). The main advantage is that the machine learning algorithms can easily fine-tune the models to the specifics of individual devices or can take uncertainty from user behaviour into account.

One step beyond, with data-driven approaches it is even not necessary any more to create a model first and then controlling it, but techniques such as reinforcement learning methods can learn a *control policy* directly. By adding a cost function (such as total energy use, greenhouse gas emissions or monetary cost), an *optimal* control policy can be found. In such a reinforcement learning method, *exploration* of the state space, and *exploitation* of the learned knowledge allow gradually improving the performance. Such reinforcement learning techniques have been successfully applied to different smart grid applications, such as controlling an electric boiler (Ruelens et al., 2018), a battery (Mbuwir et al., 2017), or a fleet of electric vehicles (Vandael et al., 2015).

However, all these data-driven applications and remote control abilities provide a number of threats as well and there is a clear need to ensure security of the application and privacy of the users.

- From a security perspective, the crucial properties are *confidentiality, integrity* and *availability*. Confidentiality can be ensured with the appropriate access control mechanisms and authentication procedures, together with encryption techniques. For integrity, adequate signature schemes and data checksum techniques can be used. Availability implies a robust information and communication technology infrastructure that performs adequately under varying load conditions. From a system's perspective, a defence-in-depth approach shall be followed, with multiple protection layers to detect and withstand intrusions, attacks or component failures.
- From a privacy perspective, when personal preferences or behaviour can be derived from the data, this can only be done when the user has given prior consent and when the European GDPR (General Data Protection Regulation) guidelines have been followed. However, a privacy-by-design approach is better and more robust than just GDPR-compliance. Such approach is based on different principles to deal with the data: minimise, hide, separate, aggregate, inform, control, enforce and demonstrate (Hoepman, 2014). As an example, it is not necessary to gather 15-min load profiles to only provide monthly billing, or it is not relevant what your gender or age is, in order to make a contract with an electricity supplier, as long as you are an adult that is legally allowed to sign a contract.

Evidently, some data (e.g., related to outages and power quality) are necessary to ensure the security of supply and to operate the grid smoothly, while other data are needed for commercial processes (e.g., consumed electricity per time-of-use period for billing).

According to the CEER (Council of European Energy Regulators), a lot needs still to be done to have the cybersecurity and privacy protection in the smart energy grids at the correct level (CEER, 2018). Especially if data are collected at finer granularity, larger resolution and more often, privacy gets at stake. It is e.g., possible to derive customer habits (like building occupancy or appliance use) from fifteen-minute electricity consumption data, or to derive more detailed info such as appliance settings from sub-second resolution apparent power measurements (Labeeuw & Deconinck, 2011).

For instance, data-driven approaches and machine learning techniques are often deployed to characterise the power flexibility from residential customers for demand response applications as they are able to consider multiple types of uncertainty, and such techniques lead to a better local flexibility characterisation and usage forecasting than approaches based on physical models (Vázquez-Canteli & Nagy, 2019). However, in order to engage fully the residential end-user in this story it is of the utmost importance to guarantee their privacy, and compliance with Europe's GDPR regulation is only a first step into this. It is our firm conviction that keeping all data at the customer's side, and hence not transferring it nor centralising it e.g., at an aggregator, is a clear advantage and will decrease customer's suspicion and allow more end user engagement.

If the data is kept locally and not transferred to multiple centralised actors, the individual has much more control about the use of the data. Of course, it assumes that some data processing is done locally, when a central actor requires information. An example of such local data processing is provided by the calculation of the total injection and grid off-take over a billing period, which needs to be calculated by the smart meter instead of forwarding the full load profiles at a 15-min resolution to a data operator. Another example is the local calculation of the available flexibility from residential appliances for demand response to be used by an aggregator, rather than sending the raw data to this aggregator.

Also, in different other sectors (such as sharing data on multiple social networks), there have been proposals to keep data local, and not to provide all data to the central databases of the social networks providers. OpenPDS (Personal Data Storage) is a clear example of this (Montjoye et al., 2014): it keeps all personal data in a local storage, and provides only 'SafeAnswers' to the allowed third parties requesting access. From a privacy perspective, this gives much more control to the user, who can select which data to share with whom, or to easily delete particular data which is no longer relevant.

In a similar way, for smart grid applications, it might be useful to look into techniques related to *computing on encrypted data*, which has been studied well in the domain of cryptography. Indeed, many applications require data to remain private (healthcare data, financial data, etc.), although one or more external parties would need to compute a specific function on this data for research, socio-economic or commercial purposes. (Here, calculates the aggregated flexibility from a cluster of residential customers without revealing flexibility data from individuals.) Multiple approaches for computing on encrypted data have been proposed in the literature. Most relevant in the context of smart grid applications are homomorphic encryption and secure multi-party computation.

- Although homomorphic encryption was conceptualized by Rivest, Adleman and Dertouzos more than 40 years ago already (Rivest et al., 1978), the construction of such a scheme that can compute 'complex' operations on encrypted data was an open problem until Gentry presented the first fully homomorphic encryption scheme using ideal lattices in 2009 (Gentry, 2009). However, this scheme and its variants were extremely slow and did not provide a practical solution. Therefore, research has focused on a more practical variant: *somewhat homomorphic encryption*, which is significantly more efficient in practice and can be used when the algorithm applied to the encrypted data is known in advance. Their main limitation is that they only allow performing a limited number of computational steps on the encrypted data.
- Multi-party computation allows multiple mutually distrustful parties to compute any function on their private inputs without disclosing those inputs to each other. Security in multi-party computation comes at a cost of efficiency. Every computation requires information exchange among the parties, which can become expensive for non-linear operations with a large multiplicative depth. Additionally, it depends on a non-collusion assumption: at least one party has to remain honest. Furthermore, the larger number of parties involved, the slower the protocol is, as more data needs to be transmitted among the parties. In recent years, practical implementation of general-purpose multi-party computation protocols have appeared and have been successfully applied in well-defined use cases, such as electricity load forecasting.

In the SNIPPET project (Montakhabi et al., 2020; SNIPPET—Secure & Privacy-friendly Peer-to-peer Electricity Trading", 2019), a secure and privacy-friendly peer-to-peer energy trading system is being developed, based on such techniques for computing on embedded data.

Besides security and privacy in a smart grid context, one also needs to consider customer protection. For instance, several types of smart meters allow for a remote connection and disconnection of customers. However, it is not acceptable that a supplier would disconnect a non-paying customer and leave him without electricity in the cold and dark during winter. In Belgium, e.g., every customer must still be supplied with at least 6A (yielding approximately 1.4 kW at 230 V) by the distribution system operator, even if all suppliers dropped him because of not paying the bills. Such regulatory framework is needed in order to protect the rights of the customers (Deconinck et al., 2011). Other customers might by apathetic or not interested in the data-driven approaches; also these customers have the right for a good power quality and a balance has to be found between sharing the minimal data for a minimal functionality with more engaged approaches that provide more features. A good overview of the context of how European policy makers deal with energy poverty is found in (Bouzarovski, 2017). Other examples of customer protection rights include the requirement to be able to check in its home at the smart meter the registers that were used for billing—they have priority over the transmitted values to the premises of the data operator.

However, the legal frameworks, at regional, national or supranational level for dealing with this customer protection, are not always up to date with the latest evolutions in smart grids technology.

Conclusion: Power to the People

In order to achieve a decarbonised energy system, changes are needed at all levels. At the transmission grids, one needs to reach out towards global grids, with power outlets on the sea for non-country specific generation; the distribution grid level needs to become an active and smart grid, rather than a passive infrastructure; the final customers will be organised in local households or in energy communities together with neighbours; independent industrial customers will make a microgrid with neighbouring companies, etc. This implies decentralised control besides the classical centralised control paradigms, combining the benefits of both approaches.

Energy communities can play a key role in decentralised control approaches. To get involvement of such communities in this energy transition, local participation is needed, so that the citizens and end users can be aware of the impact of their energy-related actions on the environment and on climate. They need to have a relationship with the energy assets and have a feeling of impact. However, the energy system has typically been organised in a top-down fashion, with centralised approaches and little active control, resulting in passive, unaware and uninterested customers. Smart grids have the potential to put active customers and consumer engagement as one of the cornerstones of a more intelligent energy infrastructure, which can be organised differently, in a more distributed way.

It is our firm conviction that the energy infrastructure needs to be prepared for a paradigm shift from centralised top-down control to distributed bottom-up control and completely decentralised peer-to peer approaches. Indeed, in different niches decentralised approaches have been used successfully (decoupled microgrids, peer-to-peer networks, etc.). The different examples in this chapter have explored how decentralised approaches can fit the future energy system, what are the related opportunities and threats, and how it can finally empower people for better engaging in the energy transition.

Acknowledgements This work has been partially supported by KU Leuven project C24/16/018, FWO project S007619N (SNIPPET), and VLAIO Flux50 projects HBC.2018.0527 (ROLECS) and HBC.2019.0073 (PrivateFlex).

References

Almasalma, H., et al. (2018). Experimental validation of peer-to-peer distributed voltage control system. *Energies, 11*(5), 1304–1325. https://doi.org/10.3390/en11051304

Almasalma, H., Engels, J., & Deconinck, G. (2017). Dual-decomposition-based peer-to-peer voltage control for distribution networks. *CIRED—Open Access Processing Journal, 2017*(1), 1718–1721. https://doi.org/10.1049/oap-cired.2017.0282

Almasalma, H., Claeys, S., & Deconinck, G. (2019). Peer-to-peer-based integrated grid voltage support function for smart photovoltaic inverters. *Applied Energy, 239*, 1037–1048. https://doi.org/10.1016/J.APENERGY.2019.01.249

Atrias. (2019). "Atrias Market Model UMIG 6.5.1.16". https://www.atrias.be/UK/Pages/Publications_UMIG65.aspx. Accessed July 11, 2019.

Bouzarovski, S. (2017). *Energy poverty: (Dis) assembling Europe's infrastructural divide*. Palgrave Macmillan.

CEER. (2018). "Cybersecurity report on Europe's electricity and gas sectors (C18-CS-44–04)," [Online]. Available: https://www.ceer.eu/documents/104400/-/-/684d4504-b53e-aa46-c7ca-949a3d296124

De Brabandere, K., Vanthournout, K., Driesen, J., Deconinck, G., & Belmans, R. (2007, June). "Control of microgrids". In *2007 IEEE power engineering society general meeting*, (pp. 1–7). https://doi.org/10.1109/PES.2007.386042

de Montjoye, Y.-A., Shmueli, E., Wang, S. S., & Pentland, A. S. (2014). OpenPDS: Protecting the privacy of metadata through safeanswers. *PLoS ONE, 9*(7), e98790. https://doi.org/10.1371/journal.pone.0098790

Deconinck, G., & Thoelen, K. (2019, August). "Lessons from 10 years of demand response research: Smart energy for customers?." *IEEE System Man, Cybernetic Magazine, 5*(3):21–30. https://doi.org/10.1109/MSMC.2019.2920160

Deconinck, G., & Vankrunkelsven, F. (2020, November). "Digitalised, decentralised power infrastructures challenge blockchains." *Proceedings Institution of Civil Engineers—Smart Infrastructure and Construction*, 1–12. https://doi.org/10.1680/jsmic.20.00013

Deconinck, G., & Vanthournout, K. (2009). Agora: A semantic overlay network. *International Journal Critical Infrastructures, 5*(1/2), 175. https://doi.org/10.1504/IJCIS.2009.022855

Deconinck, G., Delvaux, B., De Craemer, K., Qiu, Z., & Belmans, R. (2011). "Smart meters from the angles of consumer protection and public service obligations." In *16th International conference on intelligent system applications to power systems (ISAP)*. https://doi.org/10.1109/ISAP.2011.6082207

ENTSOE. "Electricity balancing." https://www.entsoe.eu/network_codes/eb/. Accessed July 11, 2019.

European Parliament. (2019). *2029/944/EU Directive of the European parliament and of the council on common rules for the internal market for electricity and amending directive 2012/27/EU (recast)*.

Ferreboeuf, H. (2019). "Lean ICT—towards digital sobriety (The Shift Project)". [Online]. Available: https://theshiftproject.org

Gentry, C. (2009). "Fully homomorphic encryption using ideal lattices." *Proceedings 41st ACM Symposium on Theory of Computing (STOC 2009), 9*(2009):169–178.

GOP2P (2021). Global observatory on peer-to-peer, community self-consumption and transactive energy models (IEA - UsersTCP). https://userstcp.org/annex/peer-to-peer-energy-trading/. Accessed April 20, 2021.

Hoepman, J.-H. (2014). *Privacy design strategies*. (pp. 446–459). Springer.

International Electrotechnical Commission. (2019). "IEC 61850:2019 SER Communication networks and systems for power utility automation". https://webstore.iec.ch/publication/6028. Accessed July 11, 2019.

Labeeuw, W., & Deconinck, G. (2011). "Non-intrusive detection of high power appliances in metered data and privacy issues." In *Proceedings 25th conference on passive and low energy architecture of the energy efficiency in domestic appliances and lighting (EEDAL) conference*, (pp. 1–7).

Lavrijssen, S., & Carrillo Parra, A. (July 2017). "Radical prosumer innovations in the electricity sector and the impact on prosumer regulation." *Sustainability, 9*(7), 1207. https://doi.org/10.3390/su9071207

Mbuwir, B. V., Ruelens, F., Spiessens, F., & Deconinck, G. (2017). "Battery energy management in a microgrid using batch reinforcement learning." *Energies, 10*(11). https://doi.org/10.3390/en110111846

Mihaylov, M., Razo-Zapata, I., & Nowé, A. (2018). "NRGcoin—A blockchain-based reward mechanism for both production and consumption of renewable energy." In *Transforming climate finance and green investment with blockchains*, (pp. 111–131). Academic Press.

Montakhabi, M. et al. (2020). "New roles in peer-to-peer electricity markets: Value network analysis." In *6th IEEE international energy conference (EnergyCon-2020)*, (p. 6).

P2P-SmarTest Project. https://www.p2psmartest-h2020.eu/. Accessed July 11, 2019.

Peter, V., Paredes, J., Rosado Rivial, M., Soto Sepúlveda, E., & Hermosilla Astorga, D. A. (2019). *Blockchain meets energy—digital solutions for a decentralised and decarbonized sector*. German-Mexican energy partnership (EP) and florence school of regulation (FSR).

Pouttu, A. et al. (2017). "P2P model for distributed energy trading, grid control and ICT for local smart grids." In *EuCNC 2017—European conference on networks and communications*. https://doi.org/10.1109/EuCNC.2017.7980652

Prettico, G., Flammini, M. G., Andreadou, N., Vitiello, S., Fulli, G., & Masera, M. (2019). *Distribution system operators observatory 2018—Overview of the electricity distribution system in Europe, EUR 29615 EN*. Publications Office of the European Union.

REScoop.eu. (2019). "Q & A : What are 'citizen' and 'renewable' energy communities ?." https://www.rescoop.eu/blog/what-are-citizen-and-renewable-energy-communities. Accessed July 04 2019.

REScoop.eu. (2020). https://www.rescoop.eu/. Accessed December 15, 2020.

Rivest, R. L., Adleman, L., & Dertouzos, M. L. (1978). On data banks and privacy homomorphisms. *Foundation Security Computer, 4*(11), 169–180

Roberts, J., & Gauthier, C. (2019). "Energy communities in the draft national energy and climate plans: Encouraging but room for improvements executive summary." REScoop.eu, European University Viadrina, p. 55. [Online]. Available: https://www.rescoop.eu/blog/necps

Ruelens, F., Claessens, B. J., Quaiyum, S., De Schutter, B., Babuška, R., & Belmans, R. (2018). Reinforcement learning applied to an electric water heater: From theory to practice. *IEEE Transactions Smart Grid, 9*(4), 3792–3800. https://doi.org/10.1109/TSG.2016.2640184

Shoreh, M. H., Siano, P., Shafie-khah, M., Loia, V., & Catalão, J. P. S. (2016). A survey of industrial applications of demand response. *Electrical Power System Research, 141*, 31–49. https://doi.org/10.1016/J.EPSR.2016.07.008

Siano, P. (2014). Demand response and smart grids—A survey. *Renewable and Sustainable Energy Reviews, 30*, 461–478. https://doi.org/10.1016/J.RSER.2013.10.022

Smart Grid Coordination Group CEN-CENELEC-ETSI. (2012). "Smart grid reference architecture". [Online]. Available: ftp://ftp.cencenelec.eu/EN/EuropeanStandardization/HotTopics/SmartGrids/Reference_Architecture_final.pdf

SNIPPET—Secure and privacy-friendly peer-to-peer electricity trading. (2019). https://www.esat.kuleuven.be/cosic/project/snippet/. Accessed July 10, 2019.

Synergrid. (2017). "C1/107 Algemene technische voorschriften voor de aansluiting van een gebruiker op het LS-distributienet".

Vandael, S., Claessens, B., Hommelberg, M., Holvoet, T., & Deconinck, G. (2013). A scalable three-step approach for demand side management of plug-in hybrid vehicles. *IEEE Transactions on Smart Grid, 4*(2), 720–728. https://doi.org/10.1109/TSG.2012.2213847

Vandael, S., Claessens, B., Ernst, D., Holvoet, T., & Deconinck, G. (2015). Reinforcement learning of heuristic EV fleet charging in a day-ahead electricity market. *IEEE Transactions Smart Grid, 6*(4), 1795–1805. https://doi.org/10.1109/TSG.2015.2393059

Vanthournout, K., De Brabandere, K., Haesen, E., Van den Keybus, J., Deconinck, G., & Belmans, R. (2005). Agora: Distributed tertiary control of distributed resources. *15th Power System Computer Conference, 2005*, 7

Vázquez-Canteli, J. R., & Nagy, Z. (2019). Reinforcement learning for demand response: A review of algorithms and modeling techniques. *Applied Energy, 235*, 1072–1089. https://doi.org/10.1016/J.APENERGY.2018.11.002

Zhang, C., Wu, J., Long, C., & Cheng, M. (2017). Review of existing peer-to-peer energy trading projects. *Energy Procedia, 105*, 2563–2568. https://doi.org/10.1016/J.EGYPRO.2017.03.737

Zhang, C., Wu, J., Zhou, Y., Cheng, M., & Long, C. (2018). Peer-to-peer energy trading in a microgrid. *Applied Energy, 220*, 1–12. https://doi.org/10.1016/J.APENERGY.2018.03.010

The Institutional Design Challenge

EU Energy Policy: A Socio-Energy Perspective for an Inclusive Energy Transition

Anna Mengolini and Marcelo Masera

Abstract This chapter presents the evolution of EU energy policy, examining how concepts of inclusiveness and justice in energy have been progressively included in relevant energy policy documents. It discusses how EU energy policy has evolved to acknowledge the importance of the individual as well as the collective dimension of energy for an inclusive green transition. Recognizing the challenges linked to the translation of these concepts into concrete actions, the chapter elaborates a socio-energy system approach that can help in making visible important aspects of the energy transition that would go unrecognized in other analytical approaches that focus mainly on the technological side. There is an increasing awareness that the European Green Deal and other political initiatives for a sustainable future require not only technological change but also careful attention to the social implications of the transition. The chapter applies the proposed approach to smart metering technologies, discussing how the technology-centric view of the energy system is framed around the average consumer or early-adopter, leaving vulnerable groups and those living in energy poverty underrepresented. A socio-energy approach also challenges the predominant use of purely quantitative results such as energy or cost savings to evaluate the successfulness of initiatives tackling inclusiveness and fairness (e.g. energy poverty). Social outcomes of energy policy choices and technology arrangements need to be better investigated and accompanied by innovative ways to measure their success. The proposed socio-energy approach offers a way of including wider societal implications of the energy transition in the design of energy policies and in their implementation.

Introduction

The European Union (EU) energy policy recognizes the central role of energy consumers *"in achieving the flexibility necessary to adapt the electricity system to variable and distributed renewable electricity generation"* (Directive (EU) 2019/944,

A. Mengolini (✉) · M. Masera
Joint Research Centre of the European Commission, Petten, The Netherlands
e-mail: anna.mengolini@ec.europa.eu

© The Author(s) 2021
M. P. C. Weijnen et al. (eds.), *Shaping an Inclusive Energy Transition*,
https://doi.org/10.1007/978-3-030-74586-8_7

2019). Empowering and providing consumers with the tools to participate more actively in the energy market, will help to achieve the EU renewable energy targets[1] and enable EU citizens to benefit from the internal market for electricity.

The European consumer policy has been based on the assumption of rational-acting consumers who tend to maximize their profits and has its roots in the information paradigm; this suggests that consumers are able, willing and competent to deal with the information provided and to take informed rational decisions (Micklitz et al., 2011). In this regard, consumers are mostly viewed as individuals and the collective dimension is largely set aside. However, recent energy policy documents present a shift towards a closer attention to the collective dimension of energy. The 2019 Clean Energy for All Europeans, while reinforcing the central role of EU citizens in the energy transition, also strengthened concepts as consumers' rights, energy poverty and vulnerable consumers, and introduced the notion of "*citizen energy communities*".[2] This articulated set of positions recognizes the importance of the collective dimension of energy production and use and how this could "*help fight energy poverty*" (Directive (EU) 2019/944, 2019) and thus strive towards a more inclusive energy transition.

The shift from generation in large central installations towards decentralized production of electricity from renewable sources is transforming the European Union's energy system, requiring the development of new strategies for handling a more decentralized system composed of heterogeneous social and technological actors motivated by different interests and agendas (Mengolini, 2017). In recent years, technological innovation and the decreasing cost of technology have made new forms of consumer participation in energy production and management more accessible (Council of European Energy Regulators, 2019). Consumers have started to produce, store and consume their own energy and are able to support the operation of power grids and energy market by changing their load patterns.

EU investments in smart electricity systems research and innovation have been steadily increasing in the last ten years (Gangale et al., 2017), however these efforts have mainly focused on the testing of enabling technological solutions without specifically addressing the needs of vulnerable consumers and the wider societal aspects of an inclusive energy transition (e.g. energy poverty) (Gangale & Mengolini, 2019). Indeed, the growing interest at policy level for an inclusive energy transition has not yet been reflected in the research and innovation initiatives carried out with EU financial support and in the implementation of current energy policies (e.g., smart metering roll out). Research and innovation (R&I) projects can play a pivotal role to address and investigate the technological, regulatory, economic and social challenges of the collective dimension of energy and to speed up the transition to an inclusive energy system with individuals and communities at its heart (Gangale et al., 2020; Gangale & Mengolini, 2019; Mengolini et al., 2016). To investigate these social challenges, the shift should be towards a socio-energy approach that would make visible

---

[1] At least 32% of energy from renewable sources in the Union's gross final consumption of energy in 2030 (Directive (EU) 2018/2001, 2018).

[2] Article 12 (Directive (EU) 2019/944, 2019).

important aspects of the energy transition that go unrecognized in other analytical approaches that focus mainly on the technological and market sides.

This chapter proceeds by presenting how issues of consumer protection, citizen engagement and inclusiveness have been receiving increasing attention in EU energy policy and how EU energy policy has evolved to recognize the importance of the individual as well as the collective dimension of energy for an inclusive energy transition (Section "The Social Dimensions of EU Energy Policy: The Role of Consumers and Communities"). Section "A Framework for an Inclusive Energy Transition" introduces a socio-energy approach to the energy transition with the aim of improving the understanding of social drivers, dynamics and outcomes of energy systems change. Section "A Socio-Energy System Approach to Smart Metering Infrastructure" applies the socio-energy approach to smart metering technologies, Section "Concluding Remarks" offers some reflections and conclusions.

The Social Dimensions of EU Energy Policy: The Role of Consumers and Communities

First Steps in the EU Internal Energy Market

Progress towards a common energy policy was limited in the first decades of European integration. The progress made with the first legislative package[3] was mainly based on internal market and environmental regulations of the EU Treaties. Neither the Treaty of Amsterdam (1999) nor the Treaty of Nice (2003) brought major progress for a common energy policy.

Major advances came only in 2007 when EU heads of state and government endorsed the first EU energy action plan that resulted in the Commission's Communication *An energy policy for Europe* (COM(2007) 1, 2007) that laid down the three major challenges for European energy policy: sustainability, security of supply and competitiveness. The action plan was followed by changes in the EU legislation. The Treaty of Lisbon (signed on 13 December 2007) added a new part on energy to the Treaty on the Functioning of the European Union (TFEU), namely article 194 in Title XXI of the consolidated TFEU. The insertion of the title on energy in the Lisbon Treaty suggests a European Union's push toward a harmonized common energy policy *"in a spirit of solidarity between Member States"* (article 194 in (TFEU, 2012)), and represents an important step forward towards a common energy policy, explicitly promoting energy efficiency and energy savings as key elements.

---

[3]Directive concerning common rules for the internal market in electricity (Directive 96/92/EC, 1996) and Directive on common rules for the internal market in natural gas (Directive 98/30/EC, 1998).

An array of new legislation followed, with the 2009 *Third Energy Package*[4] representing a further step towards the improvement of the functioning of the internal energy market. It established that all EU citizens have the right to have their homes connected to energy networks and to freely choose any supplier of gas or electricity offering services in their area. Moreover, the package urged to recognise that consumers have the right to access accurate information on their consumption data and associated electricity prices. This information on the electricity costs should be provided frequently enough in order to create incentives for energy savings and behavioural change and is facilitated by the deployment of smart metering infrastructure. Such information provision could also create innovative services to effectively enable active participation of consumers in the electricity supply market. Furthermore, the third legislative package prescribes the EU Member States to define the concept of vulnerable consumers at the national level and to adopt measures to protect such consumers and to address energy poverty.

To drive forward the consumer-related issues that were included in the Third Energy Package, in 2008 the European Commission (EC) established the *Citizens' Energy Forum* (also known as the London Forum) as a regulatory platform to help deliver competitive, energy efficient and fair retail markets for consumers. The Citizens' Energy Forum brings together national consumer organisations, industry representatives, national regulators and government authorities to discuss key issues such as approaches to protecting vulnerable consumers, price transparency, switching energy suppliers, user-friendly billing, and smart metering. Issues of energy poverty, vulnerable consumers and consumer's protection were additionally examined in a Commission Staff Working Paper, *An energy policy for consumers* (SEC(2010) 1407, 2010).

To further explore the concept of vulnerable consumers and support Member States (MSs) in the implementation of the Third Energy Package, the EC established in 2011 the *Vulnerable Consumer Working Group.*

Towards a Common Energy Union Strategy: Energy Consumer at the Centre

A major step towards the definition of a common EU energy strategy came in 2015 with *the Energy Union Package—Framework for a Resilient Energy Union with a Forward Looking Climate Change Policy*. The Energy Union strategy (COM(2015) 80, 2015) placed citizens at its core and recognized that by taking ownership of

---

[4]Directive on the promotion of the use of energy from renewable sources (Directive 2009/28/EC, 2009), Directive concerning common rules for the internal market in electricity, (Directive 2009/72/EC, 2009), Regulation on conditions for access to the network for cross-border exchanges in electricity (Regulation (EC) No. 714/2009, 2009), Regulation on conditions for access to the natural gas transmission networks (Regulation (EC) No. 715/2009, 2009), Regulation establishing an Agency for the Cooperation of Energy Regulators, (Regulation (EC) No. 713/2009, 2009).

the opportunities allowed by the energy transition, they can *"benefit from new technologies to reduce their bills, participate actively in the market"* (COM(2015) 80, 2015, p. 2) and contribute to an energy transition where vulnerable consumers are protected. The strategy set out, in five interrelated policy dimensions, the goals of an energy union: energy security, solidarity and trust; a fully integrated European energy market; energy efficiency contributing to moderation of demand; decarbonising the economy, and research, innovation and competitiveness.

The ensuing Commission's Communication *"Delivering a New Deal for Energy Consumers"* (COM(2015) 339, 2015) further clarifies the role of the consumer in the energy transition. It recognises that the combination of decentralized generation with storage options and demand side flexibility *"can further enable consumers to become their own suppliers and managers for (a part of) their energy needs, becoming producers and consumers and reduce their energy bills"* (COM(2015) 339, 2015, p. 6), thus introducing the concept of consumer-producer, also termed 'prosumer'.

From Consumers to Citizens' Joint Actions: The Collective Dimension of EU Energy Policy

While the *New Deal for the Energy Consumers* recognizes that consumers increasingly participate in collective schemes and community initiatives, *"to better manage their energy consumption"*, it is only with the *Clean Energy for All Europeans* Communication (COM(2016) 860, 2016) that the collective dimension of energy is fully recognized. The Clean Energy for all Europeans Package (CEP) argues that energy transition creates new opportunities and challenges for market participants, allowing, through technological development, for new forms of consumer participation and cross-border cooperation. It further elaborates the central role that jointly acting consumers can play in the energy transition. The implementation of CEP proposals into CEP legislative acts was finalised in 2019 and it includes eight legislative acts.[5] The CEP establishes a legislative framework where *"active customer"*[6] (definition that also includes "a group of jointly acting final customers")

---

[5]Clean Energy for all Europeans Package legislative acts: Energy performance in buildings (Directive (EU) 2018/844, 2018), Renewable energy directive (Directive (EU) 2018/2001, 2018), Energy efficiency Directive (Directive (EU) 2018/2002, 2018), Regulation on the Governance of the Energy Union (Regulation (EU) 2018/1999, 2018), Regulation on the internal market for electricity (Regulation (EU) 2019/943, 2019), Electricity Directive (Directive (EU) 2019/944, 2019), Regulation on risk-preparedness in the electricity sector (Regulation (EU) 2019/941, 2019), Regulation establishing a European Union Agency for the Cooperation of Energy Regulators (Regulation (EU) 2019/942, 2019).

[6](Directive (EU) 2019/944, 2019), article 2, point 8: *'active customer'* means a final customer, or a group of jointly acting final customers, who consumes or stores electricity generated within its premises located within confined boundaries or, where permitted by a Member State, within other premises, or who sells self-generated electricity or participates in flexibility or energy efficiency

and "*jointly acting renewable self-consumers*",[7] have more opportunities to get involved in the energy transition. Communities and individuals are given the right to produce, store, consume and sell their own energy. The Electricity Directive (Directive (EU) 2019/944, 2019) and Renewable Energy Directive (Directive 2009/28/EC, 2009) provide the definitions of "*citizen energy community*"[8] and "*renewable energy community*"[9] which are both formulated as particular ways to organise collective actions around a specific energy-related activity through the community organized as legal entity. The focus shifts from the individual consumer acting in isolation to the collective dimension of energy and to how this can contribute to a more inclusive energy transition. Indeed, the Electricity Directive highlights that community-level energy represents an inclusive option for all consumers "*to have a direct stake in producing, consuming and or sharing energy between each other*" (recital 43 of (Directive (EU) 2019/944, 2019)) and in fighting energy poverty through reduced consumption and lower supply tariffs. Energy community initiatives directly involve and engage with consumers and therefore can be best suited to "*... facilitating the uptake of new technologies and consumption patterns, including smart distribution grids and demand response, in an integrated manner*" (recital 43 of (Directive (EU) 2019/944, 2019)). The Directive further highlights that "*Energy services are fundamental to safeguarding the well-being of the Union citizens. Adequate warmth, cooling and lighting, and energy to power appliances are essential services to guarantee a decent standard of living and citizens' health. Furthermore, access to those energy services enables Union citizens to fulfil their potential and enhances social inclusion*" (recital 59 of (Directive (EU) 2019/944, 2019)). That energy transition must be fair and socially acceptable for all is confirmed in the *Fourth Report on the State of the Energy Union:* its social implications must be part of the policy process from the outset and not simply be an afterthought (COM(2019) 175, 2019). In 2018,

---

schemes, provided that those activities do not constitute its primary commercial or professional activity.

[7] (Directive (EU) 2018/2001, 2018), article 2, point 15: '*jointly acting renewables self-consumers*' means a group of at least two jointly acting renewables self-consumers in accordance with point (14) who are located in the same building or multi-apartment block.

[8] '*citizen energy community*' means a legal entity that: (a) is based on voluntary and open participation and is effectively controlled by members or shareholders that are natural persons, local authorities, including municipalities, or small enterprises; (b) has for its primary purpose to provide environmental, economic or social community benefits to its members or shareholders or to the local areas where it operates rather than to generate financial profits; and (c) may engage in generation, including from renewable sources, distribution, supply, consumption, aggregation, energy storage, energy efficiency services or charging services for electric vehicles or provide other energy services to its members or shareholders. (Directive (EU) 2019/944, 2019), article 2, point 11.

[9] '*renewable energy community*' means a legal entity: (a) which, in accordance with the applicable national law, is based on open and voluntary participation, is autonomous, and is effectively controlled by shareholders or members that are located in the proximity of the renewable energy projects that are owned and developed by that legal entity; (b) the shareholders or members of which are natural persons, SMEs or local authorities, including municipalities; (c) the primary purpose of which is to provide environmental, economic or social community benefits for its shareholders or members or for the local areas where it operates, rather than financial profits. (Directive (EU) 2018/2001, 2018), article 2, point 16.

the EC, as part of its policy efforts to address energy poverty across EU countries, launched the *EU Energy Poverty Observatory* (EPOV) with the mission to engender transformational change in knowledge about the extent of energy poverty in Europe, and innovative policies and practices to combat it.

Along the same line, the recent *European Green Deal* presents a forward-looking strategy establishing the goal that no energy consumer should be *left-behind*. It emphasises the involvement of *"local communities in working towards a more sustainable future"* and the need to further *"empower regional and local communities, including energy communities"* (COM(2019) 640, 2019, pp. 18, 21). It also advocates for a socially just transition where the risk of energy poverty must be addressed and citizens and workers most vulnerable to the energy transition must be protected.

The Twin EU Energy and Digital Transitions

In the communication *Shaping Europe's digital future* the EC recognizes that the twin challenge of a green and digital transformation has to happen together in order for Europe to lead the transition to a healthy planet and a new digital world (COM(2020) 67, 2020). The energy sector has been an early adopter of digital technologies, using them to facilitate grid management and operation. It is argued that in the next decades digital technologies will enable more connected, intelligent, efficient, reliable and sustainable energy systems (IEA, 2017). To this end, digitalization of the energy sector should be adopted along the whole value chain, from production, to distribution, consumption and management of energy. For example, smart metering systems, by providing feedback on electricity consumption, enable the consumers to monitor and manage their energy use. However, smart metering technologies implementation (and digital technologies in general) should be accompanied by an assessment of the associated societal implications to guarantee an early identification of the challenges and opportunities that the use of digital technologies and other innovative solutions can present for EU consumers' living conditions. It is therefore of paramount importance to ensure that digital development policies avoid amplifying existing inequalities and leaving vulnerable groups behind, such as those on low incomes, tenants living in multi-storey buildings and those who are digitally excluded. This requires a socio-energy approach to the energy transition to guarantee a comprehensive view of its social dimension.

The Social Dimension of EU: Energy as a Fundamental Right

The reflection paper on *The Social Dimension of Europe* warns that *"economic and technological change may result in new patterns of inequality, with a persistent risk of poverty coinciding with new forms of exclusion"*. It calls for a cohesive society that guarantees an inclusive growth and social justice (COM(2017) 206, 2017). The reflection paper paved the way for the publication of the *European Pillar of Social Rights* (European Union, 2017) that set out 20 key principles to support fair and well-functioning labour markets and welfare systems. In particular, one of the key principles is access to essential services; it recognizes that everyone has the right to access essential services of good quality, including water, sanitation, energy, transport, financial services and digital communications. Support for access to such services shall be available for those in need. Already in 2012, the *Charter of Fundamental Rights* of the EU recognized (Chapter IV, Solidarity) access to services of economic interests and consumer protection as fundamental rights (European Union, 2012). More recently, the reflection paper *Towards a sustainable Europe by 2030* (COM(2019)22, 2019) states that a transition to a low-carbon, climate-neutral, resource-efficient and biodiverse economy needs to be innovative, green, inclusive and socially just, leaving no one behind and in full compliance with the United Nations 2030 Agenda and the 17 Sustainable Development Goals.

The COVID-19 crisis came at a time when EU climate and energy policies were experiencing a new thrust with overall policy frameworks targeting carbon neutrality (e.g. the European Green Deal) and countries dealing with the implementation of climate and energy law frameworks at national level. The crisis has shown that the right to energy and energy services are essential ingredients for an inclusive energy transition (ENGAGER, 2020) and that, therefore, these new policy frameworks should be defended given their importance for structural and societal beneficial changes of the energy system (Steffen et al., 2020). A just transition that safeguards social inclusiveness and enables a fair recovery of the EU economy after COVID-19 becomes an imperative in order to maintain the political acceptability of a climate neutrality goal (IEA, 2020). In this respect, the Recovery Plan Communication *Europe's moment: Repair and Prepare for the Next Generation* advocates the need for a fair and inclusive recovery that must address disparities and inequalities either exposed or exacerbated in the crisis ((COM(2020) 456, 2020) while proceeding with the twin green and digital transitions toward a fairer and more resilient society.

The Way Forward: Energy Democracy and Energy Justice

The evolution of the main EU energy policy documents and initiatives is summarized in Fig. 1. With the onset of the Energy Union Package there has been an acceleration in the number of documents addressing the social implications of the energy transition. Although concepts of energy justice and energy democracy are not as such included in the EU policy documents analysed, one can argue that they are at the core of the EU (energy) transition. Energy justice has recently emerged as a new cross-cutting social science research agenda seeking to apply justice principles to the energy field by questioning "*the ways in which benefits and ills are distributed, remediated and victims are recognized*" (Jenkins et al., 2016). Energy democracy, on the other hand, aims at greater citizen involvement and control in the energy systems (van Veelen

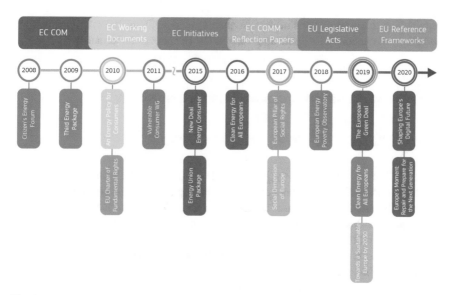

Fig. 1 EU Energy policy key documents; *EC COMM*: non-binding legal instruments that include policy evaluations, commentary or explanations of action-programmes or brief outlines on future policies or arrangements concerning details of current policy; *EC working documents*: they are geared towards providing information on certain policies, programmes and legislative proposals or in support of current policies; *EC working groups*: working groups whose mandate is to support Commission's work on specific topics; *EC COMM reflection papers*: documents outlining the view of the EC on key topics that will define the coming years; *EU legislative acts*: legally-binding acts of the European Union, such as directives and regulations; *EU reference frameworks* for social rights: framework jointly proclaimed by the European Parliament, the Council and the Commission

& van der Horst, 2018). Energy justice and energy democracy are closely related and can be considered a translation of the democracy principle and the rule of law to the energy field that are the foundations of the European Union and are at the service of a just society (Vitéz & Lavrijssen, 2020). The recognition of the collective dimension of energy enhances the role of the consumers in the transition of the energy system where they assume an active role, not only as consumers or users, but also as 'energy citizens' actively involved in shaping policies in the area of energy (Vesnic-Alujevic et al., 2016). Greater citizen involvement is at the heart of the energy democracy principle (van Veelen & van der Horst, 2018; Vitéz & Lavrijssen, 2020). As Devine-Wright suggests, 'energy citizens' is an alternative view of the public where "*the potential for actions is framed by notions of equitable rights and responsibilities across society for dealing with the consequences of energy consumption*" (Devine-Wright, 2007, p. 71). Energy citizenship contrasts the social and psychological detachment of the public from energy systems that is embedded within centralized systems. In contrast to the past view of energy consumer, for whom energy was simply a good to be expended in pursuit of personal goals, the energy citizen engages with energy as a meaningful part of their practices and is better understood in a community context. This view of an EU energy citizen with equitable rights and responsibilities in shaping and defining the energy transition should be at the core of reliable and transparent EU energy governance.

A Framework for an Inclusive Energy Transition

A Socio-Energy Approach: The Dimensions

While policies have recurrently indicated the ambition to take into consideration socio-technical issues, the translation of those aspirations into concrete actions is particularly challenging. A case in point is the Electricity Directive (Directive (EU) 2019/944, 2019), which highlights the need for an integrated approach to energy transition with no further details on how this can be achieved. Energy policy is indeed a problem of "socio-energy system design" (Miller et al., 2015). A more comprehensive approach should enable an active participation of society in the political discussions and in the elaboration and assessment of the related programmes and projects. There is an increasing awareness that the European Green Deal and other political initiatives for a sustainable future require not only technological change but also changes in consumption and social practices (Strand et al., 2021). In this context we argue in favour of a socio-energy system approach (Miller et al., 2015) that makes visible important aspects of the energy transition that go unrecognized in other analytical approaches that focus on the technological side. In applying this perspective, one should investigate the *social processes* that stimulate and manage the energy transformation, inquiring about the choices and behaviours of the social agents involved in the energy transition (business managers, policy makers, consumers…); furthermore, one should investigate the *social changes* that accompany shifts in energy technologies and that reshape social practices, values, relationships and institutions and the *social outcomes* that flow from the operation and organization of new energy systems.

In this light, and inspired by (Miller et al., 2013), we propose a framework for the deliberation of the social characteristics of energy policies that takes into consideration three dimensions: social processes, social changes and social outcomes.

Social Processes

Social processes are represented by the actions and decision making processes of various social actors. The energy transformation is the result of series of choices made by these actors (e.g. citizens, energy companies, public institutions, energy consumers, industrial users, etc.)

One fundamental *social process* in the energy transition is the communication from public authorities and energy operators to consumers regarding their plans for the deployment of assets and regulation of the retail market. The need to have structured dialogues with all stakeholders is a stronghold of all climate and energy policies. These dialogues might moderate the inevitable asymmetric nature of information access and agency capacity in any given society. The information about the issues at hand is recognised as depending on a clear political lead, and as the main enabler for the engagement of the citizens concerned (COM(2015) 339, 2015). One clear case in point is provided by the German policy for the so-called Energiewende (BMWI-BMU, 2010). It foresaw the participation of societal actors (such as citizens at large, but also NGOs, think-tanks, and even foundations linked to political parties) at different stages, looking for broad social consensus. For instance, there was, since the coming into force of the Renewable Energy Sources Act (BMWI, Renewable Energy Sources, 2017), wide dissemination of information on the feed-in tariffs that would support the installation of renewable energy generation by households (e.g. PV panels). The German Federal Ministry for Economic Affairs and Energy set up a web portal for all information regarding the Energiewende,[10] and produced the *Energiewende Direkt* newsletter that provides facts and background information on renewable energy in Germany.[11] On the other hand, there were criticisms on the lack of transparency on the parallel increase in electricity prices resulting from the feed-in tariffs. This increment has been of 22% of the average monthly electricity bill in 2016 (Ecologic Institute, 2016). In addition, the rules governing the payment for renewable power were revised, triggering uncertainty for investments made by citizens, as these are more exposed to volatility in the market.

Social Changes

The link between the social actors and the energy resource is mediated by a wide set of instruments of very different nature: from smart meters and new billing arrangements, to demand side management and differentiated tariffs, from the setting of energy communities to self-consumption. In this light, *social changes* are concerned with modifications in the appropriation, acceptance and use by the social actors of the new energy technologies, systems, services and market structures. *Social changes* are for example modifications in behaviours, especially regarding the adoption of technologies, the investment in devices, the acceptance of regulations and measures, and the variation of energy consumption patterns (Steg et al., 2018). In the last years there have been numerous studies on these behavioural changes, including the indirect and spill over effects of those changes affecting other activities relevant to energy, climate and the environment. All these social changes might be unwanted, happening as the mere result of the advancement of policy, technology and market factors, or be planned and intended as the product of an institutionalised socio-technical regime, established by authorities or operators. This distinction is fundamental for distinguishing and assessing different social dynamics.

---

[10]www.bmwi.de/EN/Topics/Energy/energytransition.html.

[11]https://www.bmwi-energiewende.de/EWD/Navigation/EN/Home/home.html.

Social Outcomes

Social outcomes flow from the operation of the emerging energy systems and may create or reinforce existing inequalities. The complex interactions of policies, technologies, and economic and social elements will oftentimes generate unforeseeable and undesirable consequences. Some of these consequences can emerge in a short period and might be easily adjusted as required, but others will cause consequences only materialising in the mid-term, so ingrained that they might be difficult to be remedied (i.e. subsidies to investment in households that expand economic disparities). The criteria used for evaluating those outcomes might change with time (e.g. how to judge the fairness of energy prices). Central to the evaluation of outcomes is the existence of multiple and contrasting standards for judging what is acceptable, and of multiple objectives that might be in contradiction with each other (e.g. energy security vs energy sustainability). From this it follows that there is no way for optimising the energy transition, and that all major decisions with society-wide impacts will require social debate and political ruling.

Critical Aspects of the Energy Transformation

These three dimensions of a socio-energy approach intersect with three critical aspects of the energy transformation that a socio-energy perspective should address: *infrastructure, knowledge* and *governance* (Miller et al., 2013).

As part of the socio-technical assessment one should examine how the decisions regarding *energy infrastructures*, which are relatively hidden from public scrutiny, impact social arrangements. Whose *knowledge* counts? Who knows about energy systems? How and what do they know? What *governance* should be put in place to implement an inclusive energy transition that will not generate negative social impacts? To answer these questions we need a socio-energy system view of the energy transition that makes all relevant aspects visible so they can be taken into due account in the governance mechanisms.

The various social actors have different levels of insight and awareness regarding the energy system. As an infrastructure, the energy system underpins and determines the social use of energy. How much of that infrastructure is perceptible by the general public? People's perceptions of all elements in the value chain are often inaccurate, from the energy source to energy consumption of devices, from the impact of technologies to the role of the energy grids, from the factors affecting the price of energy to the externalities. The situation of deficiency in perception can be better managed in local projects, where the involvement of policy makers and practitioners with the civil society actors can be fostered. Several examples demonstrate how participatory processes can help on this front (ENLARGE, 2018). However, what would be feasible at the local level, is daunting when the number of actors dramatically increases such as in vast regions. Some analytic frameworks defend for this reason a community setting for debating energy developments, but it is apparent that not all energy issues (and mainly those relating to the transmission/transport and whole-sale market) can be reduced to the local community level. This can be a source of conflicts, and merit special consideration in the governance arrangement (Brisbois, 2020; Veuma & Bauknecht, 2019).

The level of influence on the decisions affecting the energy system will be deter-mined by the *knowledge* of its structure and functioning. It is implausible to request all citizens to have a full understanding of energy systems, but it is relevant to reflect on the knowledge that might be needed by the general public.

The distance between scientific and technical knowledge on the various topics of interest, and the basic knowledge that the population at large might acquire needs to be acknowledged. In a world of rapidly changing industrial and research products, accompanied by a multitude of assessments claiming disparate and often contradic-tory results, it is hard to determine a common and solid epistemic basic reference and to organise social processes based on pure technical evidence. The presentation of the same energy choices is frequently obfuscated by knowledge mixed with ideolog-ical positions and political disputes. There is no direct and unequivocal link between more knowledge and behavioural choices (Frederiks et al., 2015; Steg et al., 2015), but knowledge affects the awareness on the matters at issue and the understanding of the potential choices by individuals and social groups.

Finally, a socio-energy approach needs *governance and justice* (Mundaca et al., 2018) (i.e. how authorities at different levels define the rules, regulations, institutions and administrative instruments affecting the infrastructure investments, the market and the rights of the social actors). As the roles of the various actors change, justice in the social processes, social changes and social outcomes is seen as a crucial factor, with participatory approaches as theme of active research (Halbe et al., 2020). As the energy infrastructure and market are organised at different planes, multi-level governance is necessary. Participative governance is recognised as central for getting the engagement of the citizens. Several open questions remain on which participatory processes can be effective beyond local communities. The extension of participatory approaches beyond this appears to be challenging.

A Socio-Energy System Approach to Smart Metering Infrastructure

The assumption of the EU energy legislation is that consumers and energy communi-ties will be active players in the energy markets, producing, consuming, selling and storing energy. There is a wealth of research that has studied consumers engagement in energy (Cseres, 2018; Gangale et al., 2013; Lavrijssen, 2014, 2017); however, not all consumers are the same. There is the need to look not only at an 'average' consumer, but also at different consumer groups in order to address the risk that certain consumers can be excluded from or be impacted negatively by the energy transition (BEUC, 2019). With this in mind, we apply the socio-energy approach discussed in the previous section to smart metering technologies.

The deployment of smart metering systems in Europe is driven by EU legislation that views smart metering infrastructure as a tool to both enhance competition in retail markets and foster energy efficiency. Moreover, smart metering technologies are considered as key enablers to realising the full potential of renewable energy integration and for the active involvement of consumers and communities in the energy transition (Bugden & Stedman, 2019). However, the potential for smart metering systems per se to trigger consumer engagement and behavioural changes is rather limited. The information on consumption provided by smart meters needs to be accompanied by a motivation to conserve, which may be provided by other instruments like financial incentives, goal setting or personal commitment (Vasiljevska et al., 2016). Smart meters can be considered as the interface between the consumers and the energy utilities and they enable consumer to interact with retailers through energy contracts. Energy contracts are becoming increasingly complex with different types of retail prices and degree of complexity concerning the associated technology. This represents a risk for consumers that may take the wrong decisions in the choice of energy contracts and services (Lavrijssen, 2017). To explore the complexity of this interaction (Mengolini, 2017; Vasiljevska et al., 2016) have developed an agent-based model that analyses the diffusion patterns of energy services (represented by energy contracts) and associated switching rate among contracts. Figure. 2 illustrates in a simplified way the interactions between the social actors and the electricity infrastructure. In Fig. 2 the consumer interacts with the electricity supplier and with the social network (community). The electricity supplier communicates with the consumer through electricity contracts, each characterized by a different type of end-user service defined in the contract and enabled by smart metering. Based on the information included in the contracts (linked to the kind of service offered, e.g.: indirect feedback, time of use pricing, home automation, …) and the interaction with the social network, the consumer will adapt and change its behaviour. The behavioural change will have an impact on the social actors and on the electricity network (social outcomes) (Vasiljevska et al., 2016). In this simplified representation of (a part of) the electricity system we can identify the three social dimensions of the socio-energy approach proposed.

Social process: institutional decisions to roll out smart meters, consumer's decision to adopt the infrastructure (voluntary or compulsory), consumer's choice of energy contracts, consumer's interaction with the social network;

Social change: consumer's acceptance of smart meters, consumer understanding of the value of the technology, consumer's behavioural change through peer interaction;

Social outcomes: consumer's achievement of own goals (financial, comfort, environmental) creates benefit for electricity network management, but can also create or reinforce existing inequalities.

The example highlights how the technology-centric view of the energy system is framed around the average consumer or early-adopter, leaving vulnerable groups and those living in energy poverty underrepresented (Rowlands & Stephen, 2016).

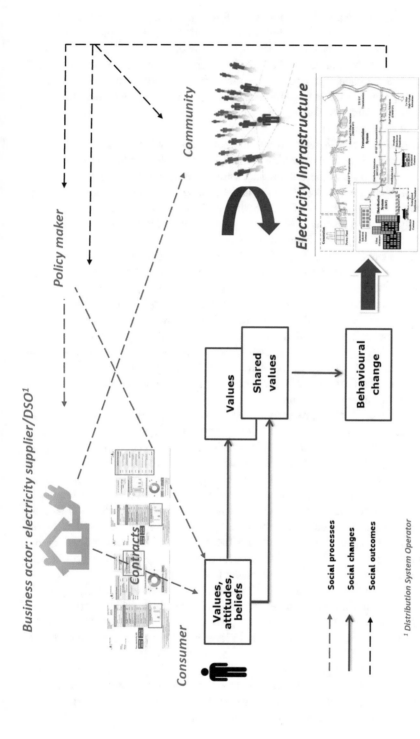

Fig. 2 Consumer in the electricity sector. Adapted from Mengolini (2017)

Wider societal implications are not at the forefront of initial considerations. The focus is mainly on technology and how it can positively transform the conditions of the average consumer. The smart home discourse revolves around "the stereotypical nuclear family in a neighbourhood of detached homes full of modern conveniences" ((Rowlands & Stephen, 2016, p. 8). The government incentive schemes designed to stimulate a greener, climate friendly energy system, may have a significant distributional effect in terms of income transfer from all taxpayers to a relatively privileged segment of the population.

If we take into consideration the vulnerable households that may be at risk of energy poverty, time of use pricing (which is one of the options offered in the contracts) could pose numerous problems for social housing occupants who may be less able than the average person to adapt to time-of-use pricing without potential damage to health and welfare. Another problem is that present arrangements concerning social assistance payments and rent subsidies do not recognize the potential adverse impacts of time-of-use pricing (Gilbert, 2006).

While the current view represents a shift from the techno-centric discourse since it includes the consumer, it does, however, not question how digital technologies will impact the society at large (governance). What does this representation say about the vulnerable consumers? What does it say on issues of energy poverty? The COVID 19 crisis has demonstrated the critical role of energy in daily lives where energy deprivation means being unable to engage with society, socially, economically and politically. While factors leading to energy poverty are multiple (low incomes, high bills, bad quality of houses), the pandemic amplifies the need to understand energy poverty better (ENGAGER, 2020) and calls for a wider appreciation of the social outcomes of the energy transition.

The deficiencies of the present view of the energy system are also reflected in the EU R&I effort for smart electrification. A recent analysis of EU R&I projects (Gangale & Mengolini, 2019) that test approaches to fighting energy poverty suggests that the growing interest in energy poverty at policy level has not yet been reflected in the research and innovation initiatives carried out with EU financial support. Many projects analysed in the report pursue multiple objectives, such as contributing to the EU energy and climate targets and alleviating energy poverty. Such objectives complement each other but often compete for priority and resources. The report suggests that more projects with a clearer focus on energy poverty and vulnerable consumers and on the wider societal aspects of the energy transition, would help to improve understanding of this phenomenon and to identify effective solutions to address it. Another interesting observation that emerges from this recent analysis and that is relevant for a socio-energy perspective is that for projects tackling energy poverty, results calculated in terms of energy or cost savings are not always a good measure of the success of the initiative. In local situations of high energy poverty, households can decide to reinvest part of the savings into higher living comfort. In these cases, the unchanged or even higher energy consumption reported after the implementation of the project activities is a sign that the project was successful in mitigating energy poverty. Future research should investigate other indicators to measure the success of the initiative, tailored to different segments of the vulnerable consumers' population (e.g. increased comfort of living, health and well-being, added market value of the property, etc.). The difficulty of shifting from a technology-centric view to a socio-energy framework clearly emerges from this analysis.

Concluding Remarks

This chapter has presented the evolution of the EU energy policy and other EU initiatives related to energy, showing how the concepts of inclusiveness and justice in the energy transition have been progressively included in relevant energy policy documents. However, the translation of these concepts into concrete actions is challenging and not yet reflected in practice. This challenge also emerges from the analysis of recent research and innovation projects in the field of energy digitalization. As a tool to address this challenge, this chapter has presented a socio-energy system approach to the energy transition. This approach allows the identification of aspects that in general go unrecognized in other analytical approaches that focus mainly on the technological side. The socio-energy system approach is applied to smart metering technologies that are viewed by EU legislation as key enablers for realising the full potential of renewable energy integration and for the active involvement of consumers and communities in the energy transition. The implementation of a socio-energy approach helps in understanding how the technology-centric view of the energy system (in the present case, smart metering deployment) is framed around the average consumer or early-adopter, leaving vulnerable groups and those living in energy poverty underrepresented. A socio-energy approach also challenges the predominant use of purely quantitative results such as energy or cost savings to judge the successfulness of initiatives tackling inclusiveness and fairness (e.g. energy poverty). Social outcomes of energy policy choices and technology arrangements need to be better investigated and should be accompanied by innovative ways to measure their success. The proposed socio-energy approach offers a way of including wider societal implications of the energy transition in the design of energy policies and in their implementation.

References

BMWI. (2017). *Renewable energy sources.* Retrieved from BMWI: https://www.bmwi.de/Redakt ion/EN/Dossier/renewable-energy.html

BMWI-BMU. (2010). *Energy concept for an environmentally sound, reliable and affordable energy supply.* Federal Ministry of Economics and Technology

Brisbois, M. C. (2020). Decentralised energy, decentralised accountability? Lessons on how to govern decentralised electricity transitions from multi-level natural resource governance. *Global Transitions,* 16–25.

Bugden, D., & Stedman, R. (2019). A synthetic view of acceptance and engagement with smart meters in the United States. *Energy Research & Social Science,* 137–145.

COM(2007) 1. (2007). *An energy policy for Europe.* Communication from the Commission to the European Council and the European Parliament. European Commission.

COM(2015) 339. (2015). *Delivering a new deal for energy consumers.* Communication from the Commission to the European Parliament, the Council, the European Economic and Social Committee and the Committee of the Regions. European Commission.

COM(2015) 80. (2015). *Energy union package. A framework strategy for a resilient energy union with a forward-looking climate change policy.* Communication from the Commission to the European Parliament, the Council, th European Economic and Social Committee anad the Committee of the Regions and the European Investment Bank. European Commission.

COM(2016) 860. (2016). *Clean energy for all Europeans.* Communication from the Commission to the European Parliament, the Council, the European Economic and Social Committee, the Committee of the Regions and the European Investment Bank. European Commission.

COM(2017) 206. (2017). *Reflection paper on the social dimension of Europe.* European Commission.

COM(2019) 175. (2019). *Fourth report on the State of the Energy Union.* Report from the Commission to the European Parliamanet. European Commission.

COM(2019) 640. (2019). *The European green deal.* Communication from the Commision. European Commission.

COM(2019)22. (2019). *Towards a sustainable Europe by 2030.* European Commission.

COM(2020) 456. (2020). *Europe's moment: Repair and prepare for the next generation.* Communication from the Commission to the European Parliament, the Council, the European Economic and Social Committee and the Committee of the Regions. European Commission.

COM(2020) 67. (2020). *Shaping Europe's digital future.* Communication from the Commission to the European Parliament, the Council, the European Economic and Social Committee and the Committee of the Regions. European Commission.

Council of European Energy Regulators. (2019). *Regulatory aspects of self-consumption and energy communities.* CEER.

Cseres, K. (2018). The active energy consumers in EU law. *European Journal of Risk Regulation,* 227–244.

Devine-Wright, P. (2007). Energy citizenship: Psychological aspects of evolution in sustainable energy technologies. In J. Murphy (Ed.), *Governing technology for sustainability.* Routledge.

Directive (EU) 2018/2001. (2018). *Promotion of the use of energy from renewable sources, European Parliament and Council of the European Union.* Official Journal of the European Union, L 328.

Directive (EU) 2018/2002. (2018). *Amending directive 2012/27/EU on energy efficiency.* Officila Journal of the European Union, L328.

Directive (EU) 2018/844. (2018). *Amending Directive 2010/31/EU on the energy performance of buildings and Directive 2012/27/EU on energy efficiency.* The European Parliament and the Council. Official Journal of the European Union, L156.

Directive (EU) 2019/944. (2019). *Common rules for the internal market for electricity and amending Directive 2012/27/EU.* European Parliament and Council of the European Union. Official Journal of the European Union, L 158.

Directive 2009/28/EC. (2009). *On the promotion of the use of energy from renewable sources and amending and subsequently repealing Directives 2001/77/EC and 2003/30/EC.* The European Parliament and the Council. Offical Journal of the European, L140.

Directive 2009/72/EC. (2009). *Concerning common rules for the internal market in electricity and repealing Directive 2003/54/EC.* The European Parliament and the Council. Official Journal of the European Union, L211.

Directive 96/92/EC. (1996). *Common rules for the internal market in electricity.* The European Parliament and the Council. Official Journal of the European Communities.

Directive 98/30/EC. (1998). *Common rules for the internal market in natural gas.* The European Parliament and the Council. Official Journal of the European Communities.

Ecologic Institute. (2016). *Understanding the energy transition in Germany.* https://www.ecologic.eu/13857

ENGAGER. (2020). *European energy poverty: Agenda co-creation and knowledge innovation.* Energy Poverty Action. Call for Action.

ENLARGE. (2018). *Report on participatory approaches in sustainable energy emerging from 'real life' practices.* https://www.enlarge-project.eu/wp-content/uploads/2018/11/D_2_1_2018_01_24.pdf

European Union. (2012). *Charter of fundamental rights of the European Union.* Official Journal of the European Union. European Union.

European Union. (2017). *European pillar of social rights.* European Union.

Frederiks, E. R., Stenner, K., & Hobman, E. V. (2015). Household energy use: Applying behavioural economics to understand consumer decision-making and behaviour. *Renewable and Sustainable Energy Reviews,* 1385–1394.

Gangale, F., & Mengolini, A. (2019). *Energy poverty through the lens of EU research and innovation projects.* Publication Office of the European Union.

Gangale, F., Mengolini, A., & Onyeji, I. (2013). Consumer engagement: An insight from smart grid projects in Europe. *Energy Policy,* 621–628.

Gangale, F., Mengolini, A., Marinopoulos , A., & Vasiljevska, J. (2020). *Collective actions in energy: An insight for EU reseach and innovation projects.* Publication office of the European Union.

Gangale, F., Vasiljevska, J., Covrig, C., Mengolini, A., & Fulli, G. (2017). *Smart grid projects outlook 2017. Facts, figures and trends in Europe, EUR 28614.* Publication Office of the European Union.

Gilbert, R. (2006). *Electricity metering and social housing in Ontario.* Social Housing Services Corporation.

Halbe, J., Holtz, G., & Ruutu, S. (2020). Participatory modeling for transition governance: Linking methods to process phases. *Environmental Innovation and Societal Transitions,* 60–76.

IEA. (2017). *Digitalization & energy.* International Energy Agency.

IEA. (2020). *European Union 2020.* Energy Policy Review.

Jenkins, K., McCauley, D., Heffron, R., Stephan, H., & Rehner, R. (2016). Energy justice: A conceptual review. *Energy Research & Social Science,* 174–182.

Lavrijssen, S. (2014). The different faces of the energy consumers. *Journal of Competition Law and Economics,* 257–292.

Lavrijssen, S. (2017). Power to the energy consumers. *European Energy and Environmental Law Review,* 172–187.

Mengolini, A. (2017). *Prosumer behaviour in emerging energy systems.* Politecnico di Torino.

Mengolini, A., Gangale, F., & Vasiljevska, J. (2016). Exploring community-oriented approaches in demand side management projects in Europe. *Sustainability, 8*(12).

Micklitz, H.-W., Reisch, L., & Hagen, K. (2011). An introduction to the special issue on "behavioural economics, consumer policy, and consumer law". *Journal of Consumer Policy,* 271–276.

Miller, C., Iles, A., & Jones, C. (2013). The social dimensions of energy transitions. *Science as Culture,* 135–148.

Miller, C., Richter, J., & O'Leary, J. (2015). Socio-energy systems design: A policy framework for energy transitions. *Energy Research & Social Sciences*, 29–40.

Mundaca, L., Busch, H., & Schwer, S. (2018). 'Successful' low-carbon energy transitions at the community level? An energy justice perspective. *Applied Energy*, 292–303.

Regulation (EC) No. 713/2009. (2009). *Establishing an agency for the cooperation of energy regulators*. The European Union and the Council. Brussels: Official Journal of the European Union, L211.

Regulation (EC) No. 714/2009. (2009). *On conditions for access to the network for cross-border exchanges in electricity and repealing Regulation (EC) No. 1228/2003*. The European Parliament and the Council. Official Journal of the European Union, L211.

Regulation (EC) No. 715/2009 . (2009). *On conditions for access to the natural gas transmission networks and repealing Regulation (EC) No. 1775/2005*. The European Parliament and the Council. Official Journal of the European Union, L211.

Regulation (EU) 2018/1999. (2018). *Governance of the Energy Union*. The European Parliament and the Council, The European Union and the Council. Official Journal of the European Union, L328.

Regulation (EU) 2019/941. (2019). *On risk-preparedness in the electricity sector and repealing Directive 2005/89/EC*. The European Parliament and the Council. Official Journal of the Europea Union, L158.

Regulation (EU) 2019/942. (2019). *Establishing a European Union Agency for the Cooperation of Energy Regulators*. Official Journal of the European Union, L158.

Regulation (EU) 2019/943. (2019). *On the internal market for electricity*. The European Parliament and the Council. Official Journal of the European Union, 158.

Rowlands, I., & Stephen, G. (2016). *Vulnerable households and the smart Grid ontario*. Metcalf Foundation.

SEC(2010) 1407. (2010). *An energy policy for consumers. Commission Staff Working Paper*. European Commission.

Steffen, B., Egli, F., Pahle, M., & Schmidt, T. (2020). Navigating the clean energy transition in the COVID-19 Crisis. *Joule*.

Steg, L., Perlaviciute, G., & Van der Werff, E. (2015). Understanding the human dimensions of a sustainable energy transition. *Frontiers in Psychology, 6*(805).

Steg, L., Shwom, R., & Dietz, T. (2018). What drives energy consumers? Engaging people in a sustainable energy transition. *IEEE Power and Energy Magazine, 16*(1), 20–28.

Strand, R., Kovacic, Z., Funtowicz, S., Benini, L., & Jesus, A. (2021). *Growth without economic growth*. European Environmental Agency.

TFEU. (2012, 10 26). Consolidated version of the treaty on the functioning of the European Union. *Official Journal of the European Union C326*.

van Veelen, B., & van der Horst, D. (2018). What is energy democracy? Connectin social science energy research and political theory. *Energy Research & Social Sciences*, 19–28.

Vasiljevska, J., Douw, J., Mengolini, A., & Nikolic, I. (2016). An agent-based model of electricity consumer. *Journal of Artificial Societies and Social Simulation*.

Vesnic-Alujevic, L., Breitegger, M., & Guimaraes Pereira, A. (2016). What smart grids tell about innovation narratives in the European Union: Hopes, imaginaries and policy. *Energy Research & Social Sciences*, 16–26.

Veuma, K., & Bauknecht, D. (2019). How to reach the EU renewables target by 2030? An analysis of the governance framework. *Energy Policy*, 299–307.

Vitéz, B., & Lavrijssen, S. (2020). The energy transition: Democracy, justice and good regulation of the heat market. *Energies, 13*(5), 1088.

Moving Towards Nexus Solutions to 'Energy' Problems: An Inclusive Approach

Ralitsa Petrova Hiteva

Abstract This chapter offers an innovative approach to examining how fuel poverty in one of the most affected countries in the EU: Bulgaria can be examined as part of the urban nexus of food, water, energy and the environment. Building on bodies of literature of the nexus, fuel poverty, energy transitions and energy geographies, this chapter uses the example of energy provisioning in the capital city of Bulgaria: Sofia to illustrate how a more inclusive approach to addressing fuel poverty and air pollution can be developed. The case study unpacks the urban nexus by examining three practices: urban gardening, making zimnina, and heating and energy use in the home. It illustrates how the interdependencies between the practices of urban gardening, making zimnina and domestic heating and energy use have direct implications for the energy system of provisioning and can be important vectors in the energy transition for vulnerable citizens in the city. The chapter addresses an important research gap in urban nexus literature by offering a compelling empirical account of mapping nexus interactions through the perspective of vulnerable users, focusing on low-technological ways of managing the urban nexus (rather than technologically driven integration across sectors).

Introduction

This chapter treats energy and energy problems such as fuel poverty as part of a nexus of food, water, energy and the environment. The chapter argues that a more inclusive (future) energy system could be built on recognising and taking into account the diverse and multiple linkages of energy and other systems, unfolding across technologies, practices, users and systems of provisioning. Unpacking the energy system as an element of the urban food-water-energy-environment nexus, this chapter uses the example of energy provisioning in Sofia, Bulgaria to illustrate how a more inclusive approach to addressing fuel poverty and air pollution can be developed. The proposed urban nexus approach is inclusive of vulnerable user experiences,

R. P. Hiteva (✉)
Science Policy Research Unit, University of Sussex, Brighton BN1 9SL, UK
e-mail: R.Hiteva@sussex.ac.uk

© The Author(s) 2021
M. P. C. Weijnen et al. (eds.), *Shaping an Inclusive Energy Transition*,
https://doi.org/10.1007/978-3-030-74586-8_8

environmental considerations and cross-sector interdependencies. The chapter also proposes a starting point for a new bottom-up institutional approach to addressing fuel poverty and air pollution building on existing practices. In doing so the chapter illustrates the importance and potential powerful impact for the lives of vulnerable energy users of social inclusiveness in transforming the energy system.

The case study discussed in this chapter examines the way energy is provisioned and specifically not consumed (in other words saved or preserved) in Sofia. Because of the case study focus on an urban environment and due to the nexus relationships unfolding in an urban context, the paper refers to the 'urban' nexus. The case study unpacks the urban nexus by examining three practices: urban gardening, making zimnina, and heating and energy use in the home. The chapter addresses an important research gap in urban nexus literature by offering a compelling empirical account of mapping nexus interactions through the perspective of vulnerable users, focusing on low-technological ways of managing the urban nexus (rather than technologically driven integration across sectors).

Fuel poverty in Bulgaria has had limited discussion within the broader EU energy landscape and most discussions have been about the negative impact of fuel poverty and its preconditions (Bouzarovski et al., 2011; Buzar, 2007; Kisyov, 2014; Kulinska, 2017; Lenz & Grgurev, 2017). There has been very limited discussion about the informal practices of vulnerable energy consumers as a way to address fuel poverty in Sofia and their potential to aid the development of a more inclusive energy system in Bulgaria and more responsive strategies for fuel poverty mitigation (EPOV, 2020). The focus of the discussion on Sofia is illustrative of how the coping practices and fuel poverty conditions discussed are applicable to many countries in Southern and Central Europe, carrying a transformative potential beyond the Sofia case study (Carper & Staddon, 2009; Bouzarovski et al., 2011; Lenz & Grgurev, 2017; Petrova & Prodromidou, 2019).

Despite having both one of the highest levels of fuel poverty and (seasonal) air pollution in the EU, these two problem areas are treated separately and with poor results. Responses to fuel poverty are disjointed and limited because of the under-pinning framing of the issue as lack of energy affordability at the point of use. The connection between fuel poverty and clean air needs to be reframed to break the vicious circle of fuel poverty leading to environmental pollution, disproportionately affecting vulnerable people.

Furthermore, at the time of writing, and to the best of the author and editors' knowledge this is the first account and discussion of energy provisioning, air pollution, fuel poverty and energy transitions in Sofia as part of the urban nexus of food, water, energy and the environment. In this respect the chapter offers a uniquely innovative approach building on bodies of literature on the nexus, ecologies of practices, fuel poverty, energy transitions and energy geographies. The chapter will be a suitable resource for undergraduate and postgraduate students; organisations working with and on behalf of vulnerable energy consumers; public policy professionals at local, regional and national level; energy and environmental practitioners; third sector researchers and representatives; and public and private organisations and individuals engaged in the energy transition.

Data about the urban nexus in Sofia was collected as part of the Resnexus project (2015–2019) funded by the Economic and Social Research Council in the UK (Ref: ES/N011414/1) to investigate the interdependencies between food, water, energy and the environment (the "urban nexus"). The data was collected between June and September 2017, and January and March 2018, and involved 42 observations, 54 interviews, 2 focus groups and 12 questionnaires. In some cases, this involved multiple observations during the summer (main urban gardening and zimnina-making period) and the winter months (zimnina consumption and heating and increased energy use in the home period).

This chapter proceeds by outlining the conceptual and empirical background of fuel poverty in the case study location, Sofia, introducing the concept of the nexus and the fuel poverty-air pollution nexus in the city (in Section "The Conceptual and Empirical Context of Fuel Poverty in Sofia"). Section "Systems of Provisioning and User Practices: Urban Gardening, Making Zimnina and Domestic Heating" outlines a nexus understanding of the systems of provisioning through the practices of urban gardening, making zimnina, and domestic heating and energy use. Section "Understanding the Fuel Poverty—Air Pollution Nexus in Sofia" situates the fuel poverty—air pollution nexus in Sofia within the urban nexus of the three practices. Section "Framing Inclusive Nexus Solutions for Inclusive Energy Transitions: Reflections and Conclusions" offers reflections and conclusions on how understanding of the urban nexus can underpin the development of an inclusive energy transition.

The Conceptual and Empirical Context of Fuel Poverty in Sofia

This section aims to unpack the multiple dimensions of fuel poverty as a geographic socio-technical concept and illustrate the scope, nature and extent of fuel poverty in Sofia. Building on these two areas of understanding this section then introduces the fuel poverty-air pollution nexus in Sofia.

Understanding Fuel Poverty—Useful Conceptual Frameworks

Fuel poverty and energy poverty are often used interchangeably although having slightly different meanings. Fuel poverty is commonly described in terms of affordability. One of the most used definitions describes households, persons or families spending more than twice the median on fuel, light and power. Frequently used measurements of fuel poverty include households that spend greater than 10 or 15% of their monthly income on energy services (such as heating or cooling); or households that actually spend more on energy than on food (Tirado Herrero & Urge-Vorsatz,

2012). "Severe fuel poverty" indicates spending 15–20% of household incomes on energy, i.e. between three- and four-times the median for a given year; "extreme fuel poverty," spending above 20% or greater than four times the median for a given year (Liddell et al., 2012).

Conceptions of fuel poverty have been heavily influenced by Boardman (1991, 2010) who initially argued that "fuel poverty occurs when a family is unable to afford adequate warmth because they live in an energy-inefficient home" (Boardman, 1993). Boardman argued that fuel poverty occurs when a household is unable to afford adequate energy services in their home on their present income, highlighting the importance of "consistent, defined standards of energy services, not just actual expenditure" (Boardman, 2012). O'Brien (2011) points out that fuel poverty is conditioned by household income as much as fuel prices and the energy efficiency of residential building stock.

People with incomes above the accepted poverty line may also not be able to afford to be warm, because their home is difficult or expensive to heat. Some energy services, such as heat, can only be purchased at the expense of adequate diets or going short in other ways (less frequent showers, socialising, buying medicine, etc.). Others with incomes sufficient to purchase adequate energy services may still live in cold conditions because of helplessness or fear of fuel bills (Bradshaw & Hutton, 1983). Poor quality of housing in terms of thermal efficiency, high levels of income inequality, and rapid increases in the real price of residential electricity lead to problems of affordability (Howden-Chapman et al., 2012). The link between all three is such that "raising incomes can lift a household out of poverty, but rarely out of fuel poverty" (Howden-Chapman et al., 2012).

Because of the impact fuel poverty has on meeting other everyday needs, the EU's definition includes households "whose resources (material, cultural and social) are so limited as to exclude them from the minimum acceptable way of life in the Member State to which they belong" (Moore, 2012). An inadequate supply of energy often means an inadequate supply for other basic domestic needs such as for food storage and cooking, maintenance of personal and domestic hygiene, and artificial lighting. Fuel poverty is a multi-dimensional phenomenon which depends as much on *who* it affects as *where* it takes place. It encompasses a wide variety of socio-demographic, institutional and built environments that render some households more vulnerable by virtue of their demographic circumstances, housing conditions and relationship to the state (Petrova & Prodromidou, 2019). Such findings point to the need to look beyond the triad of energy prices, incomes and energy efficiency within which energy poverty has been traditionally conceptualized (Bouzarovski, 2014; Petrova et al., 2013). Understanding of the concept is further expanded by the innovative work of Petrova and Prodromidou (2019), who enable drawing of key similarities between Greece and Sofia (Bulgaria)[1] and mobilise a more nuanced

---

[1]Fuel poverty in Bulgaria is often discussed alongside other South Eastern countries like Greece and Romania, because of the big similarities (social, economic and technical) between them.

understanding of fuel poverty in relation to the *urban* nexus,[2] by unpacking the complex social and spatial patterns that influence the emergence of fuel poverty.

Petrova and Prodromidou (2019) broaden the discussion of fuel poverty vulnerability by discussing the "new energy poor" in neighbouring Greece. A new category of fuel poverty which "cuts across traditional class boundaries, by including people of different genders and ages, as well as diverse ethnic and educational background" which, they argue, has emerged as a result of austerity measures and existing problems, "such as thinly insulated and inadequately heated homes, and built, institutional or ownership arrangements that do not allow households to improve the efficiency of the housing stock or switch towards more affordable fuels (Katsoulakos, 2011; Santamouris et al., 2014)". The 'new energy poor'[3] struggle to secure adequate domestic energy services due to the economic crisis and austerity regime, as well as path-dependent infrastructural and policy settings, across a variety of urban and peri-urban sites.

Petrova and Prodromidou (2019) also argue that "linking the intimate geographies of households (Valentine, 2008) with … geographies of people's dealings with austerity" can inspire the development of wider progressive politics (Jupp, 2016), calling for research on the links between energy, environmentality and austerity (Alejos & Paz, 2013), especially in the context of residential energy use. This progressive research and policy agenda also builds a more critical and in-depth investigation of how these relationships unfold beyond cities, and include integrative thinking about rural and suburban areas. Roy (2005) highlights the importance of the urban–rural interface as a key space for dynamic energy experimentation.

By unpacking the fuel poverty as part of the urban nexus this chapter also unfolds the ways in which the urban nexus is affecting the everyday rhythms and synchronicities of residential energy use (Walker, 2014), creating new temporal patterns embedded in the social world. Greater recognition of these patterns and the ways in which there are disrupted are urgently needed, as 'sustainable energy transitions and pathways … are mediated by unique place and context-specific conditions that exert influence on the mobilisation of resources, governance capabilities and actor-networks' (De Laurentis et al., 2016). Thus, holding a promise of identifying innovative and powerful ways for sustainable institutional change, towards a more inclusive energy transition.

Empirical Context of Fuel Poverty in Bulgaria

Fuel poverty in Bulgaria in general and in the capital city of Sofia, in particular, is an issue of key concern at the national and EU level. Compared to an average of 8% for the EU and 16% for the region, 39.2% of the Bulgarian population was unable to keep their homes warm in 2016. These figures continue to be the highest in the

---

[2]The concept of the urban nexus is defined in Section "Understanding the Nexus".

[3]The concept was first developed by Kaika (2012).

EU (Eurostat, 2020). It is believed that the real numbers of fuel poverty are much higher and could be over 50%. Thus, those living in fuel poverty are not only people belonging to traditionally vulnerable groups such as the elderly, disabled and those in long term unemployment, but include households in full time employment, or what Petrova and Prodromidou (2019) call the "new energy poor". Although, Bulgaria has not experienced the acute levels of austerity which took place in Greece over the past 5 years, the group of fuel poor and vulnerable continues to grow, albeit slowly.

Technology Issues: Inefficiencies and Lock-In

Fuel poverty in Sofia has strong links with a specific technology, district heating (DH) and the extent to which it is embedded in the built environment for many users. Although DH is not the only reason for the high number of people in fuel poverty, it disproportionately affects vulnerable households living in apartment type dwellings (EPOV, 2020). DH is the main form of heating and hot water supply in densely populated cities in Bulgaria, serving 26.5% of the Bulgarian population. About 65% of the national heat supply is produced by combined heat and power (CHP) plants in Sofia. However, DH suffers from strong path-dependency and lock-in: the DH sector was built during the 1950s and 1960s and was designed to provide a collective, subsidized heat supply without consideration for individual consumer needs. During this period the supply of raw resources was also strongly subsidized by the Bulgarian state and the former Soviet Union and heat was provided at a fixed price below the cost of production.

The DH system was poorly designed from the start and it was highly inefficient at the point of installation and did not allow reduction of supply costs. Furthermore, insufficient maintenance and investments led to gradual deterioration of the DH assets, low efficiency of operations, and poor quality of services. Large-scale disconnection of households from DH services (over 30%) took place between 1994 and 2000, and continues today. The decreasing customer base, with low collection rates (due to non-payments and heat and power thefts), the use of low-grade coal (lignite), growing popularity of and funding for energy efficiency retrofits, and poor energy governance in Bulgaria further weakened the financial condition of the DH company. These conditions play out particularly badly for customers living in apartment type panel buildings. In Sofia central heating (mainly DH) is used by around 11% of households and over 20% of the buildings are panel buildings, most of them needing renewal (680,000 buildings needed renewal by 2020), half of which were panel buildings.

Cost, Access and Billing

The increase in fuel (mainly gas and oil) prices toward world-market levels in the mid-1990s (following the collapse of favorable trading relations with the former Soviet Union) heavily impacted the cost of heat production and put the state budget under financial pressure. Because of these structural issues DH has long been one of the most expensive ways of providing heat and hot water in Sofia. Despite privatization and modernization changes since the early 1990s, problems with billing, with many customers being routinely overcharged and customers not having access to their heat or hot water meters persist. For example, household heat and hot water meters are locked behind specially designed cabinets to prevent tampering, with company representatives having the only keys.

DH heating bills are calculated on the basis of consumption in the same month of the previous year, which is adjusted by a fixed rate. Unseasonal cold weather in a previous year for example could impact bills a year later. Such jumps in bills are also compounded by yearly hikes in the price of heating. For example, the price of heating jumped by 23% in April 2018 compared to the previous year. The actual energy consumed by customers during the heating season becomes clear after the heat meters are read and balancing bills are issued. This means that many are faced with what is often referred to as "impossibly high bills" when the cold weather ends and people are using less heat on a daily basis. The complexity of heating bills and the fact that customers rarely have access to their heat and hot water meters leads to high levels of distrust among consumers. The impact is even bigger on vulnerable customers who feel any increases in the price more acutely. Ultimately, heating bills are a worry for many even beyond the heating season (October to March).

The DH company in Sofia is owned by the local municipality and is a monopoly. With the DH company in the red for over 2 decades, the price of heat increases on a yearly basis, while the company reports losses of over 20% due to poor infrastructure and theft of heat and hot water. The theft of heat and hot water is a widely spread practice in Sofia and can involve tampering with meters and radiators in individual properties (flats). The DH company measures the supplied amount of heat and hot water to the whole building, meaning that thefts inevitably mean higher cost for all other customers in the building. This means that ultimately, individual households do not have full control over their energy (heat and hot water) bill, something which affects many of their practices all year around. Often, the only way to exercise control over heat and hot water bills is by switching radiators off, using cold water and having greater control over how electricity is supplied in the household. The latter often involves the use of alternative technologies such as gas heaters and hobs, and wood burning stoves.

Still a Price to Pay Even if Disconnected

Many customers, particularly vulnerable customers wish to disconnect from DH altogether but are unable to if they live in multifamily buildings. Even if radiators are removed from individual properties they are still eligible for an amount to cover heat transference from pipes which run through the walls and through communal spaces, such as corridors and hallways.

Although retail electricity prices in Bulgaria are the lowest amongst countries in the EU (price per kWh), the high levels of fuel poverty and low levels of ability to keep adequately warm are multiple, systemic (connected to the historic way electricity and heat are provisioned in Sofia) and personal (linked to low earnings and old housing infrastructure). Another key issue is the low levels of energy efficiency performance in homes, with multi-family buildings being particularly hard to retrofit (Kulinska, 2017; Tirado Herrero & Urge-Vorsatz, 2012).

There are multiple barriers to overcoming fuel poverty, starting with low quality of life, unhealthy living conditions and inability to maintain the building stock. Many vulnerable citizens meet only their most pressing heating needs, using wood and coal, sometimes illegally traded. Government support is focused on subsidizing final energy consumption, which often increases energy consumption, rather than examining the coping practices of vulnerable users and the non-energy/indirect ways in which they seek to control their energy use.

Understanding the Nexus

'Nexus' is often used to denote two or several interlinked issues, systems, subjects etc. Although the word nexus means "to connect" (De Laurentiis et al., 2016), in the way it is used here it also refers to the interactions (inter/dependencies) between two or more elements, including the synergies, conflicts and trade-offs that arise from how they are managed. Multiple definitions and meanings of the nexus exist, some of which are overlapping (Al-Saidi & Elagib, 2017), however they all tend to have a "strong normative resonance" (Cairns & Krzywoszynska, 2016) towards better, more efficient and/or sustainable management of resources. Although different variations of the nexus exist—energy-water-food (EWF) nexus; food-energy-water (FEW) nexus; water-energy-food (WEF) nexus—the nexus approach is multi-centric. De Loe and Patterson (2017) suggest that what is paramount is "nexus thinking," as opposed to a specific strict definition of the nexus.

Recent nexus thinking has argued for the inclusion of climate change and the environment (Allouche et al., 2015; Pahl-Wostl, 2017) when considering the nexus. Wichelns (2017) argues that much of the interest in the nexus is a result of the concern of the impact of climate change on water, energy and food security, as all three resource sectors are influenced by climate change and that they, in turn, each contribute their own emissions (Rasul & Sharma, 2015). Most responses to the arising complexity from a nexus approach are addressed through modelling the nexus (i.e., computer-based modelling) (Daher et al., 2017). However, localising and contextualising the nexus is key to addressing its trade-offs (Daher et al., 2017; Simpson & Jewitt, 2019) and preparing an adequate policy response. Overall, since the increased interest in understanding the nexus, approaches which focus on mapping nexus interactions through the perspective of users, and vulnerable users in particular, have been few and far between.

In Sofia, as in many other cities around the world, local inequality is materialised as vast disparity, between groups of people, in terms of access to nutritious food, clean water and energy. Resilience then points not just to a capacity for adapting to the impacts of climate change and other forms of unsustainability, but also to the transformation of extant socio-ecological systems of provision towards producing social equality and ecological integrity. To begin with, such transformations require the recognition of complex interdependencies between the environment and the practices of food, water and energy production, distribution and consumption, as well as of the connection between the practices. Rather than promoting technical solutions based on data integration from the different sectors, a transformative nexus approach seeks to build on the experiences and practices of different communities, some of which are marginalised and vulnerable.

The food-water-energy-environment nexus emerges from interacting social and physical systems (de Grenade et al., 2016), in the case study of Sofia meaning practices and systems of water, energy and food provisioning, and the environment. The interdependencies between these heterogenous elements of the nexus are as much about how social and ecological systems relate to each other, as it is about the material way in which social and ecological systems are interconnected (Williams et al., 2018). Thinking in terms of a nexus means allowing for, taking into consideration and creating new interconnections across different political scales (individual practices, cities and national level policy) and between nature and people.

The fuel poverty-air pollution nexus in Sofia can be understood in the context of the wider urban nexus of food, water, energy and the environment. This would entail not focusing purely on the inter/dependencies of fuel poverty and air pollution, but embedding them (i.e. localising and contextualising them) in the relationships between three interrelated practices: urban gardening, making zimnina, and heating and energy use in the home. These three user practices are discussed in turn in Section "Systems of Provisioning and User Practices: Urban Gardening, Making Zimnina and Domestic Heating".

Understanding practices of energy use through the nexus is inclusive of the experiences of some of the most vulnerable practitioners, and of scales of analysis at the local, community and household levels, which are understudied in environmental management and nexus literature. Nexus analyses are often conducted at regional or national levels (due to the availability of data or national-level policy goals or metrics) (Miralles-Wilhelm, 2016) with smaller, more localised scales missing in most nexus discussions (Prasad et al., 2012). Common nexus approaches tend to focus on integration between systems, often through technological innovation such as smart technologies, side lining or failing to engage with low tech solutions and practices. This chapter's focus on socio-technical and ecological interconnections can also be thought of as inclusive of the agency of ecological processes, such as climate change.

Understanding the Fuel Poverty-Air Pollution Nexus in Sofia

Fuel poverty and air pollution in Sofia are tightly interlinked, and their interdependencies are at the heart of the urban nexus. The nature of these interdependencies is unpacked in this section.

Sofia is one of several most polluted European cities in South-Eastern Europe. The air pollution with particulate matter is to a large extent due to the use of wood and coal as a source of residential heating, and a direct result of the low purchasing power of the population. The multidimensional nature of the link between fuel poverty and air quality is yet to be unpacked and incorporated in current policies and measures. Fuel poverty and poor air quality are long-standing issues in the EU, but are yet to gain EU-wide recognition. As a result, policies for energy, environment and climate issues are rarely integrated (InventAir, 2018).

An Energy Agency Plovdiv (EAP) study (in InventAir, 2018) on the link between fuel poverty and air pollution in Sofia shows that households using wood and coal for heating live in worse housing conditions than the rest of the population. Their homes are usually brick-based, single-family, with a local heating boiler (58%) or without any heating infrastructure (35%). 57% of the homes do not have any insulation and 55% of the households use additional electrical devices for heating, and an electric boiler for domestic hot water (30%). Around 60% report monthly income per household member around 250 EUR and another 60% report monthly energy bill above 250 EUR. Eurostat reports that between 32 and 47% of the roundwood production in Bulgaria is fuelwood, and its use in old and inefficient stoves is a major contributor to air pollution in the winter (InventAir, 2018).

Air quality in Bulgaria has been decreasing for several years, with great implications for public health. There are regular recordings of concentrations of PM2.5 and PM10, which are much higher than the limits set by the EU and the World Health Organization to protect citizens' health. These concentrations affect up to 92% of the population, resulting in the highest rate of premature deaths due to air pollution in the EU (217 deaths per 100,000). This health crisis is attributed to household and ambient air pollution, and it is particularly bad in Sofia. Although this means that Bulgaria is in almost constant breach of EU air quality laws and was ordered by the EU Court of Justice in 2017 to take action to improve its air, little has been done. The office of the Mayor of Sofia has denied that poor air quality is an issue on multiple occasions (HEAL, 2018). This points to an institutional blindness and lack of willingness to act on the interconnections between fuel poverty and air pollution in the city. Rather than obscured and visible only to those who are vulnerable, the fuel poverty-air pollution nexus in Sofia is highly visible.

Like Bulgaria, Greece has seen a significant increase in air pollution, over the past 10 years (a 30% increase in the mass concentration of PM 2.5 particles and a up to five-fold increase in the concentration of wood smoke tracers) within large urban areas mostly due to the increased use of fuelwood for residential space heating (Saffari et al., 2013). This leads to situations where in the winter months big cities like Athens (in Greece) and Sofia (Bulgaria) are covered in smog for days (Petrova & Prodromidou, 2019; Hiteva, 2017).

In recent years, there has been strong political and financial support for renovating the outdated residential building stock, even in harder-to reach multifamily residential buildings. However, even in retrofitted properties, often people chose to disconnect from DH networks and turn to wood and coal heating. Low-income households using old inefficient heating equipment and/or poor-quality heating fuels are not the only ones to blame for dramatic seasonal increases in air pollution in Sofia. Vulnerable households are often disadvantaged not only by high fuel prices, poor infrastructure and often less flexibility in shifting their energy demand. They tend to have more limited information about energy use and options, as well as pollution levels and related health impacts. Social welfare systems often do not provide enough support regarding trade-offs between different domestic heating and energy systems (Hajdinjak and Asenova, 2019).

Because of the high levels of fuel poverty and seasonal vulnerability many households experience systemic issues with DH in the capital (around lack of transparency and billing) and rely on multiple energy resources, often including gas, electricity, DH and wood to meet their energy needs. Wood burning stoves coexist with electric radiators, central heating and air-conditioning. Sometimes in the same room. Often each room in a home will have an independent and different way of providing the necessary energy services. People in Sofia go to great lengths to make sure that they have control over where the heat comes from, how much they use and at what cost. This means that the links between fuel poverty and air pollution are deeply embedded and complex.

The fuel poverty-air pollution nexus is produced by the intersection and interdependencies between the often poor energy performance of the housing stock for many households (due to lack of adequate insulation; quality of building materials, poor maintenance etc.); the use of old, inefficient and poorly maintained heating systems; the use of low-quality fuel (wood, coal and briquettes; e.g. with high humidity or sulfur content) or even waste burning; and the high levels of fuel poverty and seasonal vulnerability in Sofia. Residential burning of low-quality fuel, often by inefficient heating systems, in poorly performing dwellings leads to substantial emissions of short-lived climate pollutants (SLCPs). Black carbon, for example, is part of particulate matter and is a particularly big contributor to global warming. Black carbon is one of the air pollutants which is commonly found in Sofia during the winter season. The reduction of particulate matter and nitric oxide pollution during the winter season is directly proportional to the reduction of the combustion of solid fuels (wood and coal) for domestic heating.

The increased use of fuelwood, according to Knight (2014), is because every crisis embeds memories of previous difficult periods, and the solutions to current problems usually involve techniques and practices used in the past. Wood-burning stoves are a traditional technology in Bulgaria for heat and cooking services, and many currently residing in urban environments like Sofia have access to such stoves in ancestral homes in rural areas. In many cases, as in Greece, the return of this 'archaic' system of space heating to urban areas can be accompanied by the reliance on more modern appliances such as air conditioners (Petrova & Prodromidou, 2019).

Limiting the harmful effects of domestic heating depends to a large extent on the successful implementation of energy policy, and although NGOs like EAP and EnEffect emphasize the importance of energy efficiency measures in buildings (HEAL, 2018), the role of user experiences and practices of energy use as part of interconnected systems of food-water-energy-environment has not been investigated. Existing energy efficiency programmes and schemes for alleviating fuel poverty often end up giving out grants to vulnerable households, who spend the money on purchasing low quality and inefficient fuels (like wood) and stoves (InventAir, 2018), inadvertently leading to nexus aggravations. This dominant approach to fuel poverty alleviation and the institutional blindness to the urban nexus relationships necessitates a new way of 'seeing' and addressing the fuel poverty- air pollution nexus in Sofia.

Systems of Provisioning and User Practices: Urban Gardening, Making Zimnina and Domestic Heating

In the case of Sofia, the nexus of food-water-energy-environment was studied through the interdependencies between the practices of urban gardening, making zimnina (food preparation for the winter) and domestic heating and energy use.

For the purposes of the study *urban gardening* is defined as comprising all activities related to the growing of food within and near the city (from inner city allotments and community gardens to peri-urban off-ground cultivation) or that have a functional relationship to it (for example, if any elements of the practice of urban gardening depend on a flow of resources and materials through the city). This could include cultivation in a wide range of urban spaces including windowsills, rooftops, balconies, building corridors, gardens, backyards, urban/peri-urban/rural allotments, spaces between blocks of apartments and alleyways. Urban gardening in Sofia tends to be small in size (practiced usually by individuals and households); and includes a wide range of informal activities, carried out primarily for self-consumption, and is driven by desire for self-sufficiency.

Urban gardening is an informal practice and involves a large number of people, of different ages and social status in Sofia. While some practice urban gardening for recreational purposes and as a way to get exercise, fresh air and produce their own food, for the majority of urban gardeners encountered, urban gardening is the means of getting fresh food for most of the year. This includes not only traditionally vulnerable groups—such as those on low-incomes, full time carers or disabled—but also people who are employed and who experience seasonal vulnerability such as fuel poverty. For some vulnerable participants, urban gardening is the only means to get fresh fruit and vegetables for most of the year.

Practices of urban gardening are closely interlinked with practices of *making zimnina*. Zimnina is a broad category which includes any food prepared and preserved for consumption in the winter months. This often involves the preservation of cooked food through canning using glass jars and metal tops. Making zimnina is very energy and water intensive, and often a complex endeavour, involving a multitude of transformations (boiling, frying, steaming, heating) and spaces (gardens, windowsills, balconies, spaces between and in front of blocks of flats, pavements and building hallways). Making zimnina in the city often necessitates the use of different types of energy. Most often these include a combination of gas, electricity and wood.

Making zimnina can be more labour, energy and water intensive for vulnerable groups (e.g., pensioners, single parents, low-income families, unemployed and disabled people) as they tend to wash, sterilize and cook the jars and produce at a high temperature to ensure that the zimnina lasts until the end of the winter (often more than 6 months). Many urban gardeners who are vulnerable to fuel poverty make use of collected wood and burn what is generally considered garbage (old clothes, plastic tubs and bottles) when preparing the zimnina (usually this takes place between May and September).

The interdependencies between the practices of urban gardening and making zimnina are particularly prominent for vulnerable practitioners. Making zimnina is an important way to redistribute food through the winter. Those suffering from fuel poverty will often survive through the winter months by eating and cooking zimnina which they make during the summer and autumn from produce they grow in their gardens. In fact, the only way many vulnerable people are able to make zimnina is if they are also able to grow their own fruits and vegetables. These interdependencies are illustrated by the experiences of one of the vulnerable participants in the study, Maria.[4]

Maria is a pensioner living 10 min away from Sofia's city centre. She has been making zimnina all her life. She starts preserving food for the winter as early as June. So, she peels, boils, fries, stuffs and closes jars with fruits and vegetables almost every day until the end of September to make enough food to last her until April next year. Experience has taught her to do this in a meticulous way: cleaning all the vegetables, fruits, glass jars and metal tops really well with plenty of water. Frying everything long enough and boiling the stuffed jars at a high enough temperature so that they last up to 6 months. She makes 1 litre jars of mixed vegetables, seasoning and cloves of garlic, which can be eaten cold, as a salad and warmed up on the wood burner in the room where she cooks, sleeps, washes herself and spends her days during the winter. Her practice is driven by the need to provide a full meal or even multiple meals for the day for herself, thus buying only bare necessities during the winter: "medicine, bread, soap". Everything else goes to bills, with heating and electricity being the biggest one. For Maria making zimnina is an important way to redistribute

---

[4]Maria's real name has been changed to protect her identity.

food through the winter, impossible without growing her own fruits and vegetables to do so.

Maria's experience is very similar to that of Anna,[5] another vulnerable practitioner of urban gardening and zimnina making in the study.

> I prepare the jars so that when I open one up it can last me the whole day. I put in everything, the seasoning, the salt, the vinegar, everything really, when I am stuffing the jars, so that I don't need to add anything else later…one half litre jar is enough for breakfast, lunch, and dinner. Sometimes I even have it as a snack…if it's really nice…. That way I don't need to switch the fridge on…If there is anything left I take it outside on the terrace where it's cold. … That's what I do in the winter. Otherwise I can't manage the heating bills. (Anna, 78-year-old pensioner, 2018)

As illustrated by Maria and Anna's cases urban gardening and making zimnina are closely interlinked with energy use in the household, and domestic heating in particular, the third practice of provisioning analysed here.

Energy provisioning in Sofia lacks the seamlessness that one can expect in cities. Multiple and overlapping ways of provisioning heat is a common place. Wood burning stoves coexist with electric radiators, central heating and air-conditioning, sometimes in the same room. This complexity of energy provisioning is not limited to heat, but also characterizes other energy practices. Households often have a portable gas hob, an electric hob and an oven, as well as access to a wood stove, usually in a property outside of the city. Practices of household heating and energy use in Sofia can include the burning of clothes, waste and low-quality fuels for the most vulnerable. Cooking in the summer is almost predominantly on a gas hob, while in the winter on an electric hob "to warm up the room". In the winter, another participant in the study, Lili[6] cooks at least three times a day as a way to keep bodies and the flat warmer, the practice of cooking and heating merging together in the winter. In comparison, in the summer, Lili usually cooks only once a day, getting up early in the morning when the air is still cool.

As the cases of Maria and Anna discussed above show, the way zimnina is prepared is closely linked to the use or saving of energy at the point of consumption. Vulnerable people often prepare zimnina in a way which does not necessitate cooking, heating up or refrigeration. What energy and how much energy will be available and used for the making of zimnina and its consumption can shape practices of urban gardening (what to grow, when to plant and when to pick produce); and how zimnina is prepared, stored and consumed. Therefore, we can think of the practices of urban gardening and making zimnina as means for resource (including energy) redistribution between the seasons. Many, particularly vulnerable people, in Sofia are managing the affordability and access to the energy services that they need (i.e. fuel poverty) through the practices of urban gardening and making zimnina. For many, coping with the energy and heating bills is a process which spreads over most of the year.

---

[5] Anna's real name has been changed to protect her identity.

[6] Lili's real name has been changed to protect her identity.

Understanding the Fuel Poverty—Air Pollution Nexus in Sofia

The practices of making zimnina, urban gardening and domestic heating in Sofia shape the interdependencies between fuel poverty and air pollution in the city. Gardeners making zimnina on a daily basis in the summer often use their allotments, gardens and forage in nearby parks and green spaces for wood, leaves, grass and any other discarded items to feed into the open fires needed to sterilize jars and close them so that they can last during the winter months (Fig. 1). In the summer the process of making zimnina can often be the source of air pollution in outside spaces, as makeshift metal stoves and fires are set up close to multifamily apartment buildings, parking lots, and even on the pavements and in building entrances (Fig. 1). In the winter months, many resort to the same practices of wood and discarded items gathering to feed wood burners set up in houses and apartment buildings, collecting from the same places they use in the summer for zimnina making. Many of the most vulnerable would scavenge the urban landscape collecting items left near and inside communal bins to top up wood specifically purchased for heating in the winter. Those who live in apartments and lack storage space, often use their gardens to store some of the collected materials for burning, regularly returning to their gardens in the winter months to ferry them to their homes.

At the peak of seasonal vulnerability, the trade-offs and choices faced by those in vulnerability for meeting their heating needs are driven as much by the things that are traditionally associated with fuel poverty (i.e. the price of energy, the energy performance of the building stock, etc.), as by their access to 'coping' practices. Coping practices such as making zimnina and urban gardening can help vulnerable

Fig. 1 Photographs of zimnina preparation (1) in the open in the street in Sofia and (2) at the entrance of multifamily apartment building at the centre of Sofia, taken September, 2017

consumers to manage affordability and access to heating services in the household. For many of the study's participants, access to land and means to practice urban gardening and making zimnina is a way to manage access to and affordability of heating in a way which allows them to practice 'healthier' energy use in the home, i.e. keeping DH radiators on, using wood burners in one room only and using biofuel central heating in houses, rather than lower grade fuels and materials. Households with larger gardens in the outskirts of the city tend to have biofuel boilers which they feed with pellets purchased specifically, mixed with leftovers from the garden and things that they are growing. The ability to synchronize the three practices of urban gardening, making zimnina, and energy use and heating through different parts of the year, in order to manage food, water and energy resources shapes the way environmental resources such as clean air and gas emissions in the atmosphere are managed at an individual and household level. This is illustrated by Hristo[7]'s words (a 57 years old taxi driver):

> It is a complicated system that has taken me years to work out. It is now well-polished and will be hard to replace… It is important to prepare early and to know what to do and when. I start collecting wood as early as May. A little bit here and there during the week, if I see some lying around I take it and put it in the car. I also collect at least a bag of branches every weekend in the park….it is just a minute or so away from the house so it is not hard to carry. Same with zimnina. A jar here or there every other day with things we find. Cheap plums at the market. The other day I saw a kilo or so of tomatoes that were going to be thrown out by the market seller and made it with green beans that I had growing on the terrace. Just a few handfuls, but when you add a clove of garlic and some herbs, it does just the job. Sometimes when we don't have a lot in the cupboard in the winter, we put a few potatoes on top of the wood burner to boil… it takes a bit longer but it does the job…, and add the tomatoes and green beans and mix it all up. If you throw in a bit of pickled cabbage that we make every year, that's me and the wife sorted for the day. It is tasty and we manage.

Although a few of Hristo's neighbours and friends are topping up their wood supplies by burning old clothes and plastics, Hristo has an asthma and is reluctant to do so. He is aware of the air pollution this causes but explains his next door neighbour's practices in terms of urban gardening and zimnina: "*… they are struggling more than we are because they are not planning ahead and not everyone can grow things as we do, either because they don't have space to do so or they don't know how. Our neighbours on the right, nice people…. hard working couple… but he has a mobility problem and can't bend down and turn the soil and she works all hours as a cleaner. They can't go and work the land as well. They do make zimnina but mostly from things that they buy on the market and that is expensive. Too expensive. We couldn't do it. … But they have to survive and they do what they can.*"

---

[7] Hristo's real name has been changed to protect his identity.

Framing Inclusive Nexus Solutions for Inclusive Energy Transitions: Reflections and Conclusions

The interdependencies between the practices of urban gardening, making zimnina and domestic heating and energy use have direct implications for the energy system of provisioning and can be important vectors in the energy transition for vulnerable citizens in the city. Including provisions for and addressing these interdependencies in policies, plans and incentives would give recognition and agency to vulnerable users to participate in urban energy transitions. Such an approach recognises vulnerable groups and practitioners as experts in their own practices. Understanding fuel poverty as part and parcel of the urban nexus of food, water, energy and the environment, allows examining the ways in which vulnerable people cope with it by managing resources over time (throughout the year) in an inclusive way, in terms of their experiences, knowledge and practices. This is in stark contrast with purely top-down approaches which seek to increase efficiencies via the use of smart technologies and integration across different systems of provisioning of the urban nexus through data platforms. The approach also opens up opportunities for more innovative and radical ways of addressing issues of fuel poverty.

Seeking only energy solutions to energy problems has some limitations. They limit possible solutions to energy actors, technologies and practices, and preclude any trade-offs possible within the urban nexus of food, water and energy. The nexus approach articulated here provides a detailed picture of interdependencies and the broader contexts within which fuel poverty interacts with the provision of food, water and its impact of the environment. In summary, the way zimnina is made and prepared is shaped and shapes energy access and use, and can affect how urban gardening is practiced. Energy use and energy saving dictates decisions about urban gardening and making zimnina, from the point of seeding to the point of consumption. The way zimnina is prepared is also closely linked to the use (or not use) of energy at the point of consumption. Vulnerable people do prepare zimnina in a way which does not necessitate to be cooked, heated up or refrigerated.

This approach will also offer a better understanding of the multiple and complex ways in which interdependence can lead to nexus aggravations. Making zimnina can be more energy and water intensive for vulnerable groups as they tend to wash, sterilize and cook the jars and produce at a high temperature to ensure that the zimnina lasts until the end of the winter. What energy and how much energy will be available and used for the making or zimnina and its consumption can shape practices of urban gardening (what to grow and when to plant); and how zimnina is prepared, stored and consumed. Interconnected practices of urban gardening and making zimnina can help stop practices of reverting to more air polluting heat provisioning in cities, such as burning of low-grade fuels, wood, plastics, old clothes and etc. With many other countries in South and Eastern Europe having traditional practices of zimnina making and climbing rates of fuel poverty, it is likely that the nexus interdependencies work in similar ways, making learning from the case of Sofia imperative beyond the country's borders.

The call for institutional change in addressing fuel poverty and designing place-based responses to environmental problems such as air pollution in cities is also clear. Rather than policies focused on the cost of energy over a limited and rigid amount of time, local and national authorities can use nexus thinking to map and understand the multiple ways in which the provisioning of food, water, energy and the environment break administrative and policy silos, and point to mobilisation of resources, governance capabilities and actor-networks towards more inclusive energy transitions. For a start, community discussions of energy transitions and fuel poverty in the urban environment, in cities like Sofia, should involve individuals, communities and in/formal networks involved in urban gardening, whose expertise and experience are part of inclusive strategies for supporting those living in fuel poverty. Such an inclusive and bottom-up approach would reframe vulnerable consumers as experts in resource distribution, rather than struggling to manage the resources at their disposal.

Advocating for support of practices that can be interpreted as enabling the reproduction of fuel poverty may feel counter-intuitive. However, fuel poverty in Bulgaria is so long-standing that it is 'normalised' (i.e. accepted by both people and institutions). The level of its embeddedness also indicates that existing energy-based and top-down financial solutions have failed to significantly reduce the number of people affected by it. Thinking of the urban nexus as a means of energy control and redistribution opens up opportunities for identifying means for self-management for vulnerable people and using them to support them. Learning from widespread bottom-up and informal practices can guide policy towards more inclusive and sustainable energy transitions in Sofia.

Supporting the practice of urban gardening would aid other related practices such as making zimnina, heating and using less energy and less polluting energy in the home. A nexus intervention aiming to alleviate fuel poverty in Sofia could involve providing support for urban gardening and making zimnina, which could range from access to land suitable for urban gardening in the city to tools and access to water and zimnina making facilities, rather than just money to cover the cost of energy, which ultimately goes to energy companies (the current approach). This would mean that fuel poverty support will not only take place during the winter months but for most of the year (as urban gardening can take place from March until November, and gardens in Sofia can produce vegetables such as broccoli and spinach even during the winter months). This means that charging rent or a fee for land use for the purposes of urban gardening (at a communal or individual basis) will rob those who are most vulnerable of an important coping mechanism against fuel poverty. However, connections across food, water, energy and the environment are poorly understood and politically neglected. There is currently limited recognition of the urban nexus interconnections and their impact on fuel poverty and air pollution at city and national level policy in Sofia and Bulgaria, respectively.

Rather than discussing fuel poverty and vulnerability as a form of personal failure or inadequacy the proposed approach focuses on the creative and proactive ways in which households deal with fuel poverty using the urban nexus. It showcases the skills and tactics developed by many households, allowing for the more efficient and effective use of energy in the home and the alternative ways of mobilizing existing infrastructures and resources of their environment to cope with an issue which is more systemic than individual in nature (because of the high number of people affected by fuel poverty in Sofia and the long period of time this has taken place). Rather than trying to bring everyone up to equitable and often higher levels of energy use, the approach aims to highlight the ways in which people try to operate with the means they have available, and in ways that are shaped by forces outside of their immediate control. A nexus approach can thus create a wider space for informal and shared networking, across siloes of public and private, individual and communal, and systems of provisioning.

Failure to recognize the deep-routed and complex linkages between fuel poverty and air pollution in Sofia, could also jeopardise shifting towards a more inclusive energy transition, one which does not leave vulnerable customers behind and lock them into a social stigma of being seen as air polluters, as well as fuel poor. Above all, the nexus case of Sofia highlights the dangers of institutional blindness which comes from the segregation of issues into separate policy silos and departments, which treat fuel poverty separately from air pollution.

References

Alejos, E., & Paz, M. (2013). An austerity-driven energy reform. *Spanish Economic and Financial Outlook, 2*, 51–60.

Allouche, J., Middleton, C., & Gyawali, D. (2015). Technical veil, hidden politics: Interrogating the power linkages behind the nexus. *Water Alternatives, 8*, 610–626.

Al-Saidi, M., & Elagib, A. (2017). Towards understanding the integrative approach of the water, energy and food nexus. *Science of the Total Environment, 574*, 1131–1139.

Boardman, B. (1991). *Fuel poverty: From cold homes to affordable warmth.* Belhaven.

Boardman, B. (1993). Opportunities and constraints posed by fuel poverty on policies to reduce the greenhouse effect in Britain. *Applied Energy, 44*, 185–195.

Boardman, B. (2010). *Fixing fuel poverty: Challenges and solutions.* Earthscan.

Boardman, B. (2012). Fuel poverty. *International Encyclopedia of Housing and Home*, 221–225.

Bouzarovski, S. (2014). Energy poverty in the European Union: Landscapes of vulnerability. *Wiley Interdisciplinary Reviews: Energy and Environment, 3*(3), 276–289.

Bouzarovski, S., Sarlamanov, R., & Petrova, S. (2011). *The governance of energy poverty in Southeastern Europe.* The Institut français des relations internationals.

Bradshaw, J., & Hutton, S. (1983). Social policy options and fuel poverty. *Journal of Economic Psychology, 3*, 249–266.

Buzar, S. (2007). *Energy poverty in Eastern Europe: Hidden geographies of deprivation.* Ashgate.

Cairns, R., & Krzywoszynska, A. (2016). Anatomy of a buzzword: The emergence of the 'water-energy-food' nexus in UK natural resource debates. *Environmental Science & Policy, 64*, 164–170.

Carper, M., & Staddon, C. (2009). Alternating currents: EU expansion, Bulgarian capitulation and disruptions in the electricity sector of South-east Europe. *Journal of Balkan and Near Eastern Studies, 11*(2),179–195

Daher, B., Mohtar, R. H., Lee, H., & Assi, A. (2017). Modeling the water-energy-food nexus. In P. Abdul Salam, S. Shrestha, V. Prasad Pandey, & A. K. Anal (Eds.), *Water-energy-food nexus: Principles and practices.* (pp. 55–66). Wiley.

de Grenade, R., House-Peters, L., Scott, C. A., Thapa, B., Mills-Novoa, M., Gerlak, A., et al. (2016). The nexus: Reconsidering environmental security and adaptive capacity. *Current Opinion in Environmental Sustainability, 21*, 15–21.

De Laurentiis, V., Hunt, D. V. L., & Rogers, C. D. F. (2016). Overcoming food security challenges within an energy/water/food nexus (EWFN) approach. *Sustainability, 8*, 95.

de Loe, R. C., & Patterson, J. J. (2017). Rethinking water governance: Moving beyond water-centric perspectives in a connected and changing world. *Natural Resources Journal, 57*, 75–99.

EPOV (EU Energy Poverty Observatory). (2020). *Member state report: Bulgaria.* https://www.ene rgypoverty.eu/sites/default/files/downloads/observatory-documents/20-06/extended_member_ state_report_-_bulgaria.pdf

Eurostat. (2020). *Inability to keep home adequately warm—EU-SILC survey.* https://appsso.eur ostat.ec.europa.eu/nui/show.do?lang=en&dataset=ilc_mdes01

Hajdinjak, M., & Asenova, D. (2019). Sustainable energy consumption and energy poverty: Challenges and trends in Bulgaria. In F. Fahy, G. Goggins, & C. Jensen (Eds.), *Energy demand challenges in Europe*. Cham.

HEAL. (2018). *Air pollution and health in Bulgaria facts, figures and recommendations.* Briefing. Air Quality. https://www.env-health.org/wp-content/uploads/2018/11/HEAL-Brief-Pos_AIR_ Bulgaria.pdf

Hiteva, R. (2017). *Affordability and unevenness in provisioning of water and energy in Sofia.* Resnexus blog. https://resnexus.org/blog/affordability-unevenness-provisioning-water-energy-sofia/

Howden-Chapman, P., Viggers, H., Chapman, R., O'Sullivan, K., Telfar, L. B., & Lloyd, B. (2012). Tackling cold housing and fuel poverty in New Zealand: A review of policies, research, and health impacts. *Energy Policy, 49*, 134–142.

InventAir. (2018). *InventAir: Report on the energy poverty and air quality status in the Eastern European Countries.* https://www.inventair-project.eu/images/IA-Status-report.pdf

Jupp, E. (2016). Families, policy and place in times of austerity. *Area, 49*(3), 266–272.

Kaika, M. (2012). The economic crisis seen from the everyday. *City, 16*(4), 422–430.

Katsoulakos, N. (2011). Combating energy poverty in mountainous areas through energy-saving interventions. *Mountain Research and Development, 31*, 284–292.

Kisyov, P. (2014). *Report on national situation in the field of energy poverty—Bulgaria.* REACH https://reach-energy.eu/wordpress/wp-content/uploads/2014/12/D2.2-EAP_EN.pdf

Knight, D. M. (2014). *A critical perspective on economy, modernity and temporality in contemporary Greece through the prism of energy practice.* London School of Economics and Political Science, London.

Kulinska, E. (2017). Defining energy poverty in implementing energy efficiency policy in Bulgaria. *Economic Alternatives, 4*, 671–684.

Lenz, N. V., & Grgurev, I. (2017). Assessment of energy poverty in New European Union member states: The case of Bulgaria, Croatia and Romania. *International Journal of Energy Economics and Policy, 7*(2), 1–8.

Liddell, C., Morris, C., McKenzie, S. J. P., & Rae, G. (2012). Measuring and monitoring fuel poverty in the UK: National and regional perspectives. *Energy Policy, 49*, 27–32.

Miralles-Wilhelm, F. (2016). Development and application of integrative modeling tools in support of food-energy-water nexus planning—A research agenda. *Journal of Environmental Sciences, 6*, 3–10.

Moore, R. (2012). Definitions of fuel poverty: Implications for policy. *Energy Policy, 49*, 19–26.

O'Brien, M. (2011). *Policy summary: Fuel poverty in England.* Lancet.

Pahl-Wostl, C. (2017). Governance of the water-energy-food security nexus: A multi-level coordination challenge. *Environmental Science & Policy, 92*, 356–367.

Petrova, S., & Prodromidou, A. (2019). Everyday politics of austerity: Infrastructure and vulnerability in times of crisis. *EPC: Politics and Space, 37*(8), 1380–1399.

Petrova, S., Gentile, M., Mäkinen, I. H., & Bouzarovski, S. (2013). Perceptions of thermal comfort and housing quality: exploring the microgeographies of energy poverty in Stakhanov, Ukraine. *Environment and Planning, A45*, 1240–1257.

Prasad, G., Stone, A., Hughes, A., & Stewart, T. (2012). *Towards the development of an energy-water-food security nexus based modelling framework as a policy and planning tool for South Africa.* Presented at Strategies to Overcome Poverty and Inequality Conference, University of Cape Town, Cape Town.

Rasul, G., & Sharma, B. (2015). The nexus approach to water–energy–food security: An option for adaptation to climate change. *Climate Policy, 16*, 682–702.

Roy, A. (2005). Urban informality: Toward an epistemology of planning. *Journal of the American Planning Association, 71*(2), 147–158.

Saffari, A., Daher, N., Samara, C., et al. (2013). Increased biomass burning due to the economic crisis in Greece and its adverse impact on wintertime air quality in Thessaloniki. *Environmental Science & Technology, 47*(23), 13313–13320.

Santamouris, M., Alevizos, S. M., Aslanoglou, L., et al. (2014). Freezing the poor—Indoor environmental quality in low and very low income households during the winter period in Athens. *Energy and Buildings, 70*, 61–70.

Simpson, G. B., & Jewitt, P. W. G. (2019). The development of the water-energy-food nexus as a framework for achieving resource security: A review article. *Frontiers in Environmental Science, 7*, 1–8.

Tirado Herrero, S., & Urge-Vorsatz, D. (2012). Trapped in the heat: A post-communist type of fuel poverty. *Energy Policy, 49*, 60–68.

Valentine, G. (2008). The ties that bind: Towards geographies of intimacy. *Geography Compass, 2*(6), 2097–2110.

Walker, G. (2014). The dynamics of energy demand: Change, rhythm and synchronicity. *Energy Research & Social Science, 1*, 49–55.

Wichelns, D. (2017). The water-energy-food nexus: Is the increasing attention warranted, from either a research or policy perspective? *Environmental Science & Policy, 69*, 113–123.

Williams, J., Bouzarovski, S., & Swyngedouw, E. (2018). The urban resource nexus: On the politics of relationality, water–energy infrastructure and the fallacy of integration. *Environment and Planning C: Politics and Space, 37*(4), 652–669.

Good Governance and the Regulation of the District Heating Market

Saskia Lavrijssen and Blanka Vitéz

Abstract This chapter discusses how the fundamental values of energy democracy and energy justice and the principles of good governance can play a role in developing a more consistent approach towards the regulation of the energy sector and, more in particular, in dealing with the challenges of regulating the heat transition in the Netherlands in a just way. Energy justice and energy democracy are energy specific concepts that are gaining influence when interpreting and applying the principles of good governance in the energy sector. Both concepts are based on the awareness that the energy transition is a matter for all citizens of the European Union and should not be ignored by policymakers and independent regulators. The heat transition in the Netherlands significantly impacts the position of consumers, prosumers and vulnerable customers, as an ever-larger group of consumers will be disconnected from the gas grid and will be connected to heat networks. Energy democracy and energy justice and the principles of good governance are important values that should guide policy-makers in making choices that affect consumer participation and the protection of vulnerable customers in the heat transition. It is elaborated how energy democracy and energy justice and the principles of good governance indeed can provide a useful framework within which advantages and disadvantages can be weighed of regulatory choices to be made when modernising the regulation of the heat market in a just way. In particular, there remains a lot to gain in terms of flexible regulation and supervision as well as the facilitation of consumer/prosumer participation in the Netherlands. Because it is likely that most heat consumers will remain locked in for a relatively long time in natural monopolies facilitated by older generation heat networks and the lack of alternative heating, substantive consumer-participation could yield positive results regarding community engagement in heat network management and heat supply.

This paper is an updated and extended version of B. Vitez & S. Lavrijssen, The energy transition: Democracy, justice and good regulation of the heat market. Energies, 2020 13(5), 1–24. [1088]. https://doi.org/10.3390/en13051088.

S. Lavrijssen (✉) · B. Vitéz
Tilburg University, Tilburg, The Netherlands
e-mail: S.A.C.M.Lavrijssen@tilburguniversity.edu

Introduction

The importance of heat networks in the Netherlands is growing under the influence of the energy transition, set in motion by the Paris Agreement of 2015 (United Nations, 2015). The energy transition embodies the move away from a fossil-fuel based economy in favour of a low carbon economy. The energy transition plays a pivotal role in policy decisions preventing or mitigating climate change (Lavrijssen, 2018). New forms of energy generated with sustainable sources and new ways of storing energy are being stimulated[1] (CBS, 2019). The combat against climate change also fuels the transition of the (Dutch) energy sector into a smart energy system. This transition towards a smart energy system is characterised by several elements (Boersma, 2015; ECORYS, ECN, 2014; Frontier Economics, 2015; Overlegtafel Energievoorziening, 2015). It entails a shift from centrally generated energy from fossil fuels to energy generated from more local, renewable sources, that are often volatile, like wind energy and solar energy. Energy consumers increasingly become prosumers: they not only withdraw energy from the distribution network, but also produce energy themselves and feed it into the network. On 28 June 2019, the Dutch government presented a national Climate Agreement that shows the way to reducing greenhouse gases by 49% in 2030 in the Netherlands, compared to the levels in 1990 (Klimaatakkoord, 2019). One of the identified solutions to reduce CO_2—emissions is the expansion of the heat market: the provision of heat via heat networks.

To replace fossil fuels for the heating of residential areas by heat delivered via heat networks (Klimaatakkoord, 2019), regional and local public authorities have been tasked with developing schemes to disconnect local areas (households) from the gas grid which will have a huge financial and spatial impact on the heat consumers. And while heat networks are gaining more importance in the Netherlands, so is the discussion regarding the appropriate market organisation and adequate regulation of the heat market (ACM, 2018; ECORYS, 2016; Schepers, 2009). For the purposes of this article, a heat consumer is defined as a household—i.e., a small consumer, connected to a heat network with a connection of a maximum of 100 KW.

In spite of the on-going changes in the energy sector, the legislative framework and legal safeguards currently in place are still (largely) based on the traditional market model, in which centrally managed, large-scale production units supply energy to meet user demand. From this perspective, users are viewed as passive actors rather than active players wanting to act as prosumers (Parag & Sovacool, 2016). A discrepancy between technological and legal developments in the energy sector and the assumptions of the existing regulatory framework can be identified, which can be seen as an example of regulatory disconnection that needs to be restored (Bukento, 2016; Lavrijssen, 2018). This discrepancy is also identifiable in the Dutch heat market, where old-fashioned assumptions underlie the existing regulatory design.

---

[1]In 2017, only 6.6% of energy generated in the Netherlands came from sustainable sources such as wind turbined and solar panels. In 2020, that percentage should be 14, in accordance with the Renewable Energy Directive 2009/28/EC.

For instance, the maximum price for heat is based on the price of gas (no-more-than-otherwise principle; *niet meer dan anders principe*). This principle entails that the maximum price that consumers will pay for heat is based on the price of natural gas. As the supply of natural gas is being phased out in the Netherlands, it is no longer sustainable to maintain its supply price as reference point. To spur the energy transition in the Member States, the European Union has adopted the Clean Energy Package (European Commission, 2016) to facilitate the integration of renewable energy in the electricity system and to enhance the protection and empowerment of electricity consumers and prosumers. The integration of renewable energy requires a more holistic and systematic approach towards the regulation of the energy sector, including the heat market. However, regulation of the heat market is a challenge, as it is not harmonised by EU law. Its regulation differs substantially from the regulation of the electricity and the gas sectors.

A fundamental question is whether a more coherent and transparent approach is possible and which values and principles could play a role in developing a more consistent approach towards the regulation of the energy sector and, more in particular, in dealing with the challenges of the regulation of the heat market in a just way. A just energy transition refers to the fundamental values of energy justice and energy democracy, and the importance of these two core values is increasingly being recognised for the regulation and market organisation of the energy sector (Jenkins, 2019; Lavrijssen & Vitez, 2020). This chapter discusses energy justice aims at a fair distribution of energy, starting by questioning "the ways in which benefits and ills are distributed, remediated and victims are recognized" (Heffron et al., 2016). Energy democracy, on the other hand, is aimed at the involvement of citizens in the energy sector as 'energy citizens' (Van Veelen et al., 2018).

In addition to these core values, also the principles of good market supervision and regulation are considered relevant for the regulation of the energy sector. The principles of good market supervision and regulation (good governance), help ensuring and fostering economic development (OECD, 2014). By providing a sound normative framework within which governance and regulation take their shape, the principles of good governance may also help attain a high-quality regulatory environment for the heat market. While good governance is subjective and depends on various elements (Andrews, 2008), the principles of good governance create boundaries within which good governance exists. In that way, these principles may also play a harmonising function by providing common principles on which regulation may be based, whilst leaving room for specific regulatory arrangements depending on the economic and technological characteristics of a certain market.

The above-mentioned considerations lead to the following question: *If and how can the core values of energy democracy and energy justice and the principles of good market regulation and supervision play a role in regulating the energy transition, in particular by dealing with the challenges of regulating the Dutch heat market in a just way?*

In order to answer this question, it might be worth taking note of a more established heat market. In 2017, 63% of Danish households were connected to a heat network (Danish Energy Agency, 2017). District heating plays a significant role in

Denmark's heat supply history since the late 1970s. In 1979, Denmark passed its first Heat Supply Act. Since then, Denmark has actively shaped its heat networks sector (Danish Energy Agency, 2017). A salient characteristic of the Danish heat market is the pervasive presence of citizen involvement. Denmark's long-standing practice provides an excellent opportunity to analyse how consumer participation shapes the heat market and how this relates to choices in governance and regulation.

The research principally relies on legal analysis of the current and upcoming European and national rules that are relevant for the economic governance and regulation of the Dutch heat market and networks. This is supplemented by the analysis of accompanying documents, reports, etc. to the applicable rules, such as explanatory statements from the Dutch legislator. Additionally, the legal analysis of the applicable rules draws on legal and economic theory on governance and regulation, in order to understand the legal requirements and institutional models for the regulation of the heat networks and the impact of good governance on the working of markets. The research also comprises a review of literature and case law on the concepts of energy democracy and energy justice, the principles of good governance and on heat markets. Part of the research is comparative as a comparison will be drawn between the Dutch heat market and the regulation of the Dutch electricity market, and consumer participation in Denmark will be explored. Regulation of the electricity market and consumer participation in Denmark have been chosen because of their characteristics: the electricity market has been subject to European regulation since the 1990s. Since that time, not only regulatory changes but also substantive changes in the sector have taken place—notably the transition from fossil fuels to renewable energy sources that is well under way. Denmark on the other hand, is noteworthy as it is progressive in organising its heat market in a way to stimulate consumer participation.

This chapter will firstly introduce the concepts of energy justice and energy democracy and the link with the principles of good governance and elaborate on their relevance for the heat market. Subsequently, the characteristics of the Dutch heat market will be described, and the Dutch Heat Act will be introduced. Following this comparison, the main developments to take place in the Dutch heat market will be considered and a comparison on points will be drawn between the rules applicable to the Dutch heat market and the electricity market. This will allow to identify the main economic and legal challenges, that will be assessed in the sixth section. The Danish heat market will briefly be discussed to see whether inspiration can be drawn for the regulation of the Dutch heat market regarding citizen participation. Lastly, the research question will be answered by a conclusion on how the values of energy democracy and energy justice and the principles of good governance can play a role in decision-making regarding the design of market organisation, regulation and supervision in a way that restores the regulatory disconnect in the heat market and provides for a more coherent and just approach towards energy regulation.

Role and Function of the Principles of Good Governance

The Concepts of Energy Democracy and Energy Justice

The core values of energy democracy and energy justice embrace the affordability of energy, its security of supply and the overall sustainability of the energy sector which are key values of EU energy policy (Edens, 2017). Energy justice and energy democracy can be seen as sector specific interpretations of the democracy principle and the rule of law (Lavrijssen & Vitez, 2020). The aforementioned concepts embody significant substantive values and as such, offer leads to policy makers (and independent regulators) on how to pursue the goals—or public values (Bruijn & Dicke, 2006)—of the Energy Union. By implementing these values in regulation, energy justice and energy democracy have a direct effect on regulation and its application. This is for instance foreseeable with regard to the principle of consumer participation, where they provide for more substantive interpretation of consumer participation also including financial participation and local ownership. Whereas the concepts of energy justice and democracy are relatively newer interpretations of the rule of law and the democracy principle, they are still in development (Pellegrini et al., 2020).

Energy democracy focuses on collective participation of citizens in energy projects (Van Veelen et al., 2018; Morris & Arne, 2016). Energy democracy is thus aimed at reforming the current organisation and decision-making process in the energy sector by advocating reform. Instead of a top-down approach, energy policies should be as much bottom-up as possible. The often necessarily decentralised nature of many renewable energy sources fits in well with the aims of the energy democracy concept. Smaller scale projects leave more room for citizen initiatives, citizen participation and citizen ownership and encourage community engagement (Alarcón & Chartier, 2018).

Aimed at tackling disparities in our energy system, energy justice is the counterpart of energy democracy and refers to the decision-making process for energy projects (Bickerstaff et al., 2013; LaBelle, 2017; Sovacool et al., 2017). As such, the concept of energy justice *"Seeks to apply justice principles to energy policy, energy production and systems, energy consumption, energy activism, energy security and climate change* (Heffron & McCauley, 2017; Heffron et al., 2016; Jenkins et al., 2016)." Energy justice thus also questions the existing state of affairs in the energy sector, and plays a role in formulating what should happen, from a perspective of what would be 'just' (Jenkins, 2019). This reveals that equality—and the strive for equality which is a fundamental part of the rule of law—is at the root of energy justice (Pellegrini et al., 2020). When taking equality as the starting point of energy justice, no definition has emerged as authoritative: energy justice is multi-faceted (Pellegrini et al., 2020). The versality of energy justice means that studies on energy justice are typically concerned with three fundamental forms of justice: distributive justice—*who gets what*, procedural justice—*who is involved in decision making*, and justice as recognition—*who is ignored or misrepresented in the energy system* (McCauley et al., 2013; Sovacool & Dworkin, 2015; Sovacool et al., 2015).

That the values of energy democracy and energy justice are increasingly finding their way into the governance of the energy sector is partially illustrated by the increasing attention to citizen participation in European regulation by referring to the role of the citizen energy community and the concept of active consumer. The recast Energy Directives refer to the benefits of citizens' participation, thereby starting to embrace the concept of energy democracy and energy justice. In doing so, the concepts of energy democracy and energy justice are being fleshed out from a legal viewpoint.

The significance of energy democracy and energy justice for the purposes of both regulation and for the principles of good governance in the energy sector cannot be underestimated. Moreover, while it is too early to draw any conclusions, the question might be raised whether the increasing importance of energy justice and energy democracy will lead to the development of separate principles of good energy governance.

The Principles of Good Regulation

The principles of good regulation first took the main stage in the World Bank's policy statements. In 1992, the World Bank issued a booklet '*Governance and Development*' that recognised the importance of good governance for economic development (The World Bank, 1992). In its foreword, Lewis T. Preston—the then-president of the World Bank—stated that "*efficient and accountable management by the public sector and a predictable and transparent policy framework are critical to the efficiency of markets and governments, and hence to economic development*". It is safe to say that today, accountability and transparency are still the basis for good governance. In addition, other principles are recognised for their role in shaping good governance. These principles concern independence, participation, effectiveness and efficiency.

Since the 1990s, the principles of good governance have become well-established principles in law and economics. Regardless of the fact that the exact delineation of which principles comprise 'the' principles of good governance is not set in stone, the common ground of the principles identified by various international organisations provide a good proxy for the most relevant principles[2] (Council of Europe, 2008; European Commission, 2001; OECD, 2014; Ottow, 2015).

Some of the principles of good governance have even explicitly been acknowledged in European case law, and currently play a large role on both European and national level (European Commission, 2001). Since their emergence, their use and interpretation have been adapted. The legal nature of the principles is different

---

[2]The Council of Europe for example, has identified no less than 12 principles of good governance and even awards local authorities achieving a high level of good governance. The European Commission identifies five principles in its White Paper that underpin good governance. The OECD details fundamental principles for the governance of regulators "to develop a framework for achieving good governance". These five principles are openness, participation, accountability, effectiveness and coherence.

according to their development over time. In the European Union, the principles of good governance provide a basis for legislation and regulation in the energy sector and other network sectors (Hancher et al., 2004). This basis consists of norms which, although differently coloured according to the relevant situation, provide a core of *"normative, universal values"* (Lavrijssen, 2006) which are generally reflected in legislation and practice as norms that are guaranteed. As such, several principles of good governance are embedded in European energy policies. Unlike the gas and electricity markets, the heat market is, as yet, largely unregulated by the European Union.

As explained above energy democracy and energy justice are values that colour the principles of good governance in the energy sector and are influencing regulation of the energy sector (Gonzalez et al., 2018). These sector-specific interpretations of the rule of law and the democracy principle are directing the further development and interpretation of the principles of good governance in the energy sector. In the absence of European regulation, an assessment of the regulation of the Dutch heat market is all the more relevant. This will provide an opportunity to assess where and how the principles of good governance can play a role in restoring the growing regulatory disconnect between theory and practice in the Dutch heat market. Such a 'fitness-check' can help identify ways to bring the current regulation up to speed to meet the (imminent) demands of the energy transition. Whereas many of the principles of good governance are multifaceted, having (slightly) different meanings according to the exact use, it should be pointed out that the principles are solely discussed with an eye on the requirements for the organisation of economic regulation of the heat market. Therefore, certain meanings and sub-principles are omitted[3] (Lavrijssen & Vitez, 2020; Lavrijssen & Vitez, 2015) (Fig. 1).

Independence

In early case law, the Court of Justice of the European Union stated that the national regulatory authority in charge of the application of economic regulation needs to be independent from market parties (EUR-Lex, 1991a, 1991b, 1993; Lavrijssen & Ottow, 2011). To fully prevent regulatory capture, one must go a step further (after all, not only market parties but also stakeholders (may) have their own agendas) (Ottow, 2015), and independence from all market parties—public and private—is required (Ottow, 2015). In light of Article 4(3) TEU, which contains the principle of sincere cooperation, independence from all market parties needs to be achieved to ensure an effective application of EU (competition) law (De Visser, 2009; Lavrijssen & Ottow, 2011). In order to guarantee fair competition, the principle entails that a regulatory authority should be independent from all market parties (Larouche, 2014). This can partially be guaranteed by the law itself if it provides conditions

---

[3]To make up for this absence of a full explanation of each principle, the references in this paper may be consulted for more information on the discussed principles.

Fig. 1 Energy justice and
democracy (Lavrijssen &
Vitez, 2020)

and restrictions for its application by the responsible regulatory authority. However, laws cannot regulate and predict every economic aspect of a market. Therefore, laws have to be sufficiently flexible to adjust to changing economic, environmental and social circumstances. This can be ensured by attributing a regulatory authority with a sufficient degree of discretion to act within the regulatory framework (Hancher et al., 2004).

A second aspect of independence—political independence—is not (yet) as firmly established (Hancher et al., 2004). Political independence refers to "*the degree to which [an] agency takes day-to-day decisions without the interference of politicians in terms of the offering of inducement or threats and/or the consideration of political preferences*" (Koop et al., 2018). The OECD has found that there is a need for regulators to be politically independent as this supports public confidence in the objectivity and impartiality of their decisions and effective operation thereby increasing the trust-levels in the market (OECD, 2016, 2017).

But it still remains controversial to demand from Member States to separate their regulatory authorities entirely (or even partially) from political influence (Lavrijssen & Ottow, 2011). An OECD report on the governance of regulators, recognises this (OECD, 2014). The stance of the OECD elucidates that choices that are predominantly of a political nature should be left to a Ministry. Applied to the heat market, political policy choices made by the government include decisions concerning the affordability of heat. Conversely, in order to realise such policy objectives, an independent regulatory authority must take (day-to-day) regulatory decisions independently and use different instruments autonomously, including the establishment of the methods of tariff regulation (Larouche et al., 2012).

Accountability

While independence is indispensable to guarantee objective and consistent decision-making, there is a danger that independence will lead to a regulatory authority acting beyond its mandate (Lavrijssen & Ottow, 2012). In order to curb this risk, a well-functioning accountability mechanism is required. Accountability and independence are therefore two sides of the same coin—demonstrating a constant tension between them (Lavrijssen & Ottow, 2012).

Bovens defines accountability as *"a relationship between an actor and a forum, in which the actor has an obligation to explain and to justify his or her conduct, the forum can pose questions and pass judgment, and the actor may face consequences."* (Bovens, 2006) This practical definition focuses on the process of giving account. In the organisation of economic regulation of the heat market, accountability should in the first place be directed towards the government (Larouche, 2014). This is referred to as political accountability and entails that an economic regulator renders account to a representative body (Aelen, 2014).

Political accountability expresses a possibility for democratic control, as in the end, voters give feedback on the results of the pursued policies (Aelen, 2014). This is desirable as it allows the Minister, the Parliament and, at the end of the accountability chain, the electorate (Bovens, 2006), to establish whether public interests are duly protected by a regulatory authority. This guarantees the proper functioning of an independent regulator and strengthens its independence[4] (Larouche, 2014).

Secondly, a regulatory authority also needs to give account to its stakeholders, including heat consumers, in a more direct way. This is referred to as social accountability (Larouche, 2014). Social accountability is likely to increase support for the activities of the regulatory authority. In that regard, stakeholders might discover incidents in which their interests have insufficiently been taken into account by the regulator, or the regulator has followed the wrong procedure according to a stakeholder, etc. (Lavrijssen, 2006). As a consequence, social accountability coupled with legal standing rights, gives stakeholders the chance to refer matters to the judiciary as an extra control-mechanism (Lavrijssen, 2006).

Transparency

The principle of transparency flows from the principle of democracy[5] (Aelen, 2014; Prechal & De Leeuw, 2008) whereby it pursues two different aims in the context of economic regulation. Firstly, it provides for legitimacy of a regulatory authority's

---

[4]It should be noted that also in the absence of an independent regulator, accountability of the regulator is equally important; see for the relationship between independence and accountability Lavrijssen & Ottow 2011 and Lavrijssen & Ottow 2012.

[5]Prechal and De Leeuw (2008) relate the principle of transparency not only to the principle of democracy, but also to the right to be heard and the rights of defence.

independence[6] (Aelen, 2014) and secondly, the principle of transparency contributes to the effectiveness[7] (Aelen, 2014) of economic regulation. In European (case) law, several aspects of the principle of transparency have been recognised, such as the right of access to documents[8]—also enshrined in the Public Access Regulation (EUR-Lex. Regulation (EC) No 1049/2001).

The definition given by Hancher, Larouche and Lavrijssen thoroughly denotes the requirements imposed by this principle upon economic regulators: a regulatory authority needs to be open with stakeholders about its objectives, processes, record and decisions. Moreover, authorities should explain to the citizens and the regulated firms the rationales of their decisions. Given that authorities are liable to be 'captured' (at least as far as their attention and their information is concerned) by regulated firms, the principle of transparency could even go as far as to require authorities to actively seek the involvement of other interests, in particular customers and citizens, in their activities (Hancher et al., 2004).

Participation

From the definition of the principle of transparency, a transition to the principle of participation is easily made (Lavrijssen, 2006; Addink, 2005; Mendes, 2011; Alemanno, 2013). Participation of all stakeholders is essential to benefit economic regulation. Stakeholders include heat consumers, consumer organisations, lobby groups, NGOs, etc. (Lavrijssen, 2006). This principle has been acknowledged implicitly by the Court of Justice of the European Union. In *Council v Access Info Europe* for instance, the Court notes in respect of the right of access to documents, that access to documents *"enables citizens to participate more closely in the decision-making process"* (EUR-Lex, 2013). Participation is also referred to in Article 11 TEU.

According to the European Commission, *"improved participation is likely to create more confidence in the end result"* (European Commission, 2001). Creating more confidence in the end result thus entails participation in the process leading to that result. Nevertheless, it should be noted that the predominantly soft law nature of the principle of participation has as a consequence that interested parties have limited possibilities to enforce participatory rights. In that way participation may have less impact on increasing the acceptance of the outcome of a decision making process (Alemanno, 2013). In this regard, enforceable rights of participation are better placed to increase the legitimacy of regulatory outcomes.

---

[6] According to, Aelen (2014) legitimacy is understood in the sense that the regulator may be independent, but only if it is guaranteed that the regulator will provide insight in its actions. In that way, being transparent legitimises the independence of the regulator.

[7] According to, Aelen (2014) transparency contributes to effective regulation in different ways. For example, publication of monitoring information by the regulator contributes to transferring the applicable norms to regulated parties—thereby possibly achieving a higher rate of compliance.

[8] Articles 41 and 42 of the EU Charter of Fundamental Rights.

Furthermore, a distinction could be made by type of participation: procedural participation versus substantive participation (Lavrijssen & Vitez, 2020). Procedural participation is best described as the right of stakeholders to be consulted at set points—like a (public) consultation on a draft regulation, or—for a more specific example—the ENTSO-E consultation process applicable to, *i.a.*, network codes; etc. Substantive participation on the other hand, refers to an on-going process of participation by stakeholders—as a constant and direct influence on the governance of energy projects. Here the idea is that participation of citizens (and other stakeholders) goes beyond a formal tick-the-box exercise and could also include financial participation and local ownership in energy projects. Furthermore, participation should be inclusive, meaning also vulnerable customers should be able to participate in the energy transition[9] (EUR-Lex, 2019).

This trend is also reflected in the rise of the significance of energy democracy and energy justice in the European Union (see below). For an example of substantive participation, the Renewable Energy Directive (EUR-Lex, 2009, 2018) comes to mind. The Renewable Energy Directive notes that the participation of local citizens and local authorities in renewable energy projects through renewable energy communities has resulted in substantial added value in terms of local acceptance of renewable energy and access to additional private capital. This results in local investment, more choice for consumers and greater participation by citizens in the energy transition[10] (EUR-Lex, 2018). Likewise, Article 16 of the recast Electricity Directive addresses citizen energy communities, of which voluntary participation is an important aspect[11] (EUR-Lex, 2019). Both directives require Member States to provide for an enabling regulatory framework facilitating and stimulating that citizens can participate in local renewable energy projects. Member States shall ensure that energy communities are able to access all electricity markets in a non-discriminatory manner.

Effectiveness

This principle of good regulation needs to be distinguished from another principle of effectiveness, often referred to by the Court of Justice of the European Union in its

---

[9]Directive (EU) 2019/944 of the European Parliament and of the Council of 5 June 2019 on common rules for the internal market for electricity and amending Directive 2012/27/EU, considerations 125–199. According to Article 28 each Member State shall define the concept of vulnerable customers which may refer to energy poverty and, inter alia, to the prohibition of disconnection of electricity to such customers in critical times. The concept of vulnerable customers may include income levels, the share of energy expenditure of disposable income, the energy efficiency of homes, critical dependence on electrical equipment for health reasons, age or other criteria.

[10]Directive (EU) 2018/2001, consideration no. 70.

[11]Directive (EU) 2019/944, considerations 125–199.

case law concerning the application of European law in national legal orders[12] (Aelen, 2014; EUR-Lex, 2010). The Commission states that the principle of effectiveness as a principle of good regulation entails that *"[p]olicies must be effective and timely, delivering what is needed on the basis of clear objectives, an evaluation of future impact and, where available, of past experiences. Effectiveness also depends on implementing EU policies in a proportionate manner and on taking decisions at the most appropriate level."*[13] (European Commission, 2001) This definition shows that the principle of effectiveness is non-binding, yet subject to the binding nature of the elements relating to the principles of subsidiarity and proportionality that applies to all EU action.[14] Effectiveness should act as an obligation resting upon both legislator and regulator when drafting legislation, policies and taking decisions—taking into account the principles of subsidiarity and proportionality (Aelen, 2014; OECD, 2012). The national dimension of the principle of effectiveness as a principle of good regulation, is not shaped from an 'obligation imposing' viewpoint. Rather, it starts from the viewpoint that the government serves the public interests (Aelen, 2014). This starting point leads to the interpretation of the principle of effectiveness as a requirement that public intervention must be efficient and effective (Aelen, 2014).

Efficiency

Effectiveness implies a need for efficiency, bringing us to the principle of efficiency. Whereas efficiency can be defined in multiple ways, this principle is multifaceted. A regulator that acts cost-effectively by carrying out its mandate in a way that requires the least possible input, or brings about the least possible costs, is no doubt efficient. However, verifying whether this type of efficiency has been achieved, is nigh-impossible: at the time of decision-making, most regulators will not have all relevant information to reach the most efficient outcome (Baldwin et al., 2012). Therefore, the principle of efficiency does not require absolute results, but necessitates that a regulatory authority is mindful of efficiency considerations.

For the governance of the heat market, the principle of efficiency bears most relevance as it concerns looking at a market as a whole. In order to guarantee a market that operates efficiently, intervention should only take place when a market does not operate efficiently. Only a market failure—*i.e.*, monopoly, information asymmetry, etc.—justifies intervention, and only in so far as it remedies the perceived market failure (Den Hertog, 2010). Viewed in this way, the principle of efficiency is strongly linked to the principle of subsidiarity as found in Article 5 TEU (Portuese, 2011).

---

[12] See ECJ C-246/09 par. 25, and Aelen (2014), p. 153. In that context, the principles of effectiveness implies that national procedural laws may not render the exercise of rights flowing from EU law "practically impossible or excessively difficult".

[13] ,European Commission (2001), European governance—a white paper, COM(2001) 428 final p. 7.

[14] See Article 5 TEU. The principles of subsidiarity and proportionality apply to (the use of) all EU competences.

Subsidiarity in the European Union means that powers shared between the European Union and Member States are executed at the lowest appropriate level of governance. A similar rationale may apply to efficient market organisation: efficiency means that governance is limited to facilitating necessary interventions by the best-placed actors in order to attain a well-functioning market. In addition to these economic grounds, a market failure in the energy sector may come in the form of public values that will not be adhered to when left to the 'free market'. The definition of an efficient European energy sector includes certain values that society wants to safeguard. These values (such as affordability, security of supply and sustainability) are equally relevant in deciding whether intervention is necessary (Prosser, 2006).

Characteristics of the Dutch Heat Market

Heat Networks

In the Netherlands, heat networks come in many forms and types. There are approximately 18 large district heating networks and approximately 100 smaller heat networks. The combined heat delivered by these heat networks (to households, buildings and greenhouses) is 22 petajoules (PJ) per year. Approximately 400,000 households are connected to district heating (Huygen et al., 2019). Smaller types of heat networks include block heating (*blokverwarming*), where one heat source supplies heat to an apartment complex or to a block of houses. District heating (*stadsverwarming*) is slightly larger in scale and usually involves the use of residual heat that stems from companies or that is released during the generation of electricity by power plants (Rijksoverheid, 2021) Heat/cold storage on the other hand, is used to store heat or cold in the ground. The following main distinctions can be made (ECORYS, 2016) [15]:

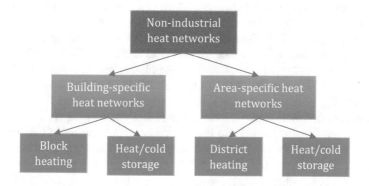

_____

[15]Table made based on data from, ECORYS (2016) p. 28, 29, 30.

Not only the form and type of heat networks differ significantly from each other. The same applies to ownership and size. Most heat networks are privately owned by companies in the Netherlands and heat is supplied by a vertically integrated heat supplier. In 2015, 40% of all heat consumers were connected to a heat network from one of only five players[16] (CBS, 2017; ECORYS, 2016). These five players are responsible for one third of all heat supplied to heat consumers and provide heat to entire cities or regions in the Netherlands. In addition, housing corporations and homeowners' associations each account for a quarter of the heat supply to heat consumers. The latter two typically supply heat via block heating and are relatively small-scale (ECORYS, 2016).

Heat networks may also be classified in 'generations' (see table below). It may be no surprise that newer heat networks are technologically more advanced than older networks. First generation heat networks for example, are steam based.[17] Newer generations of heat networks support increasingly lower water temperatures, still allowing for the release of heat (Buffa et al., 2019). Instead of steam, second generation heat networks supply pressurised hot water, at temperatures above 100 °C. These heat networks were built between 1930 and 1970 (Lund et al., 2014). Most heat networks in use today are third generation heat networks that distribute heat via water at temperatures below 100 °C. Due to the lower water temperature, third generation heat networks can be fuelled by a greater variety of heat sources (not limited to fossil fuels) compared to the first two generations, thereby ensuring a more efficient heat use. This opened the way to the use of biomass and waste as heat sources, sometimes even supplemented by solar and geothermal heat (Lund et al., 2014).

Technological developments and the possibility to integrate renewable heat sources, such as geothermal heat, into the heat grid, open the way to the use of smart energy systems with integrated systems for electricity, heating and cooling (Huygen et al., 2020; Lund et al., 2018). The development of 4th and 5th generation district heating includes the supply of increasingly more energy efficient buildings with space heating and warm water, while reducing losses in heat grids (Lund et al., 2018). To enable the use of these new heat networks, many buildings and production processes must be made suitable for this type of heat supply—a costly process (Lund et al., 2018; Hoogervorst, 2017; Huygen et al., 2020). Fourth generation heat networks go a step further than third generation networks and are well-equipped to supply heat to modern, energy-efficient, buildings with lower heat demands (due to better insulation, etc.) (Huygen et al., 2020; Lund et al., 2018). The lower water temperature (around 65 °C) means that transport losses are significantly reduced. Fifth generation heat networks distribute water at a close to ambient ground temperature. Transport losses are significantly limited and installation costs are lowered. Since the water temperature is relatively low, it may also become easier to add parties to the network (Buffa et al., 2019; Lavrijssen & Vitez, 2020). In the case of fifth generation networks, it would even be possible to add to the network small consumers acting as prosumers

---

[16]These players are Eneco, Nuon, Ennatuurlijk, Stadsverwarming Purmerend, and HVC.

[17]Fueled by coal or excess steam from industry.

by supplying excess heat to the heat grid (Buffa et al., 2019; Lavrijssen & Vitez, 2020). Fourth, and especially fifth, generation heat networks are not yet prevailing in the Netherlands (Natuur en Milieu, 2018).

Characteristic	Heat network		
	3rd generation	4th generation	5th generation
Introduction	Appr. 1970s	Appr. 2010s	Appr. 2020
Water temperature	Below 100 °C	Around 65 °C	Between 8 and 25 °C
Heat source	Oil, biomass, waste incineration, solar thermal, geothermal, excess heat from industry	Higher share of renewable sources possible	All-renewable heat sources possible
Heat storage	Yes	Yes	Yes, also cold storage possible
Energy meters	Metering and monitoring	Smart	Smart
Cooling	No	Possible	Yes—and building can feed excess heat into network (heat-sharing)
Ideal for	District heating	Well-insulated, energy-efficient buildings	Well-insulated, energy-efficient buildings with heating and cooling needs

Features of the Market

Taking into account current technical limitations, older generation heat networks are plagued by considerable transport losses. To mitigate this, heat networks in the Netherlands tend to be local and decentralised. Another feature shaping the market is that the location of a heat source cannot in all instances be chosen freely. Combined with the fact that most current heat networks in use in the Netherlands are prone to transport losses, significant limits are posed upon the location of heat networks (Tieben & Van Benthem, 2018).

Furthermore, in case the production of heat is linked to other processes (for example in the case of the use of residual heat), the predictability and reliability of heat production are not straightforward. After all, the production of heat is then dependent upon another process. In these cases, investing in heat production proves to be more complicated compared to investing in single production processes (Tieben & Van Benthem, 2018). In addition, heat networks are closed systems, meaning that the water in the network is pumped around and does not leave the networks—only the heat is delivered to consumers. This makes it generally more challenging to add parties to the loop as this may make the system more vulnerable to loss of heat, quality, etc. Lastly, because heat-demand fluctuates, is seasonal and difficult to store, auxiliary heat sources have to be ready to deliver heat to the grid if demand outgrows supply (Tieben & Van Benthem, 2018).

As a result of these features, the heat market is comprised of local, natural monopolies. Most heat consumers have no alternative heat sources due to a lack of alternative connections. This situation is likely to remain, because as of the 1st of July 2018, no gas grid connection is supposed to be provided for newly built houses in the Netherlands.[18] Instead, new dwellings should be connected to a heat network or provided with a heat pump or other means of heating. Hence, heat consumers are locked into long-term heat solutions whilst having no viable alternative heat source to switch to. This raises the question whether and to what extent heat consumers should be protected, and if so, what this protection should entail (Lavrijssen et al., 2013). In some cases, heat consumers wishing to terminate their heat supply contract, are prevented by law from doing so.[19] This is the case when a heat supplier can prove that it is technically impossible to stop the supply of heat to that consumer or that termination would lead to a significant disadvantage to another heat consumer.

Applicable Laws and Regulations

Dutch heat market legislation stems from before the energy transition. Consequently, changes instigated by the energy transition are generally not reflected in the applicable laws and regulations. This means that the envisioned role of heat networks in the energy transition and the legal implications this brings, are potentially not supported by the regulatory system (Zilman et al., 2018). Currently the Minister of Economic Affairs is preparing a new Heat Act (Act on Collective Heat Systems) stimulating the roll out of sustainable collective heat systems in the Netherlands (Akerboom & Huygen, 2021).

The heat market comprises the production, transport/distribution and delivery of heat (ACM, 2018; Tieben & Van Benthem, 2018). Currently, players on the heat market are not subject to (legal) unbundling requirements. This means that both vertically integrated firms and non-integrated firms may be active. Nor is there any requirement in the Netherlands that heat networks should be owned by public authorities, unlike for electricity and gas networks. Heat consumers are typically dependent upon vertically integrated suppliers, who are in charge of network management, transport/distribution and delivery.

The majority of the applicable rules are included in the Dutch Heat Act. The Dutch Heat Regulation and the Dutch Heat Decision—both based on the Dutch Heat Act—provide further specifications. On the 1st of July 2019 a revised version of the Dutch Heat Act has entered into force.[20]

---

[18] Article 10(7)(a) and (b) Dutch Gas Act.

[19] Article 3c Dutch Heat Act.

[20] Dutch Ministry of Economic Affairs and Climate Policy, Staatsblad (2018), 311.

The revised Dutch Heat Act bridges the period until the Act on Collective Heat systems will be implemented. The Act on Collective Heat Systems is currently under consultation and builds on the legislative process started in February 2019 that will lead to the adoption of an entirely new heat act, envisaged to be in force as of 1 January 2022 (Ministerie van Economische Zaken en Klimaat, 2020). The legislative process will focus on three main themes: market organisation, tariff regulation and sustainability (Ministerie van Economische Zaken en Klimaat, 2019). Hence, this is an appropriate moment to assess the current regulation of the Dutch heat market and examine opportunities to restore any regulatory disconnects between regulation and practice and to see whether there are sufficient guarantees for a just energy transition in the Dutch heat market.

The Dutch Heat Act first entered into force on the 1st of January 2014 and was preceded by a legislative process spanning approximately ten years (Eerste Kamer der Staten-Generaal, 2003). Until then and compared to the rules in place to protect other energy consumers, heat consumers were left in the cold. The introduction of the Dutch Heat Act was aimed at protecting heat consumers, who were left at the whims of monopolist heat suppliers.[21] Regulation was desired in order to balance the situation in which electricity and gas users could both benefit from a liberalised market and enjoy (some) legal protection whereas heat consumers could and did not (McGowan, 2001). It was feared that heat consumers could be charged exceedingly high prices or had to settle for unsatisfactory service from heat suppliers—without any specific remedies at hand (Dutch Parliamentary Papers, 2002).

The current Dutch Heat Act prohibits the supply of heat to consumers without a license. This prohibition does not apply to suppliers (i) serving at most ten heat consumers concurrently; (ii) who do not exceed a supply of 10,000 gigajoules per year; or (iii) who are lessor or owner of the building for which the heat is supplied.[22] Non-licensees do not have to adhere to section 2.2. of the Dutch Heat Act. This section contains rules regarding the grant and withdrawal of a licence, services offered by the licensee, and accounting requirements. The scope of the applicability of the Dutch Heat Act thus depends on whether a supplier is required to be licensed to supply heat.[23] This partially explains the lack of precise information on the number of heat consumers connected to heat networks.

The revised Dutch Heat Act further limits the scope of the act by way of a new Article 1a. This Article stipulates that the Dutch Heat Act does not apply to heat suppliers that—in short—are also a lessor or homeowner's association and supply heat to their lessees or members.[24] Reason for this change is that the previous provisions imposed an administrative burden on, in particular, block heating provided by

---

[21] Of course, competition law has a role to fulfil here too. However, protecting heat consumers via competition law has certain setbacks like the fact that it is carried out ex post and whereas a breach of competition rules can be fined, this does not indemnify heat consumers who suffered from the breach.

[22] Article 9 Dutch Heat Act.

[23] It follows from Article 9(2) Dutch Heat Act that generally blocks heating as well as very small players are exempted from adhering to section 2.2. of the Dutch Heat Act.

[24] Article 1a Dutch Heat Act. With the exception of parts of Articles 8 and 8a.

homeowner's associations who qualified as large-scale suppliers under the previous Dutch Heat Act (Dutch Parliamentary Papers, 2016). Several additional concerns are also addressed now. For example, the Dutch Heat Act offered (and still offers) protection by way of a price-cap to heat consumers with a connection of a maximum of 100 KW. Homeowners' associations and housing corporations buying heat and then reselling it to their members or lessees were previously not protected by this price-cap because their own connections exceed 100 KW. Despite that homeowner's associations and housing corporations did not benefit from the price-cap, in their capacity as 'large-scale supplier', they were obliged to offer this protection to their members and lessees. This could result in a discrepancy between the (uncapped) purchasing and (capped) reselling price. By excluding homeowner's associations and housing corporations from the scope of the revised Dutch Heat Act, these problems do no longer exist. Heat consumers purchasing heat from their landlord, are no longer protected by the Dutch Heat Act, but instead by tenancy law. Also, the new Act on Collective Heat Systems, which is currently under consultation, exempts homeowners' associations and lessors from certain requirements and subjects them to a lighter regulatory regime.

The Dutch Heat Act requires 'large-scale' heat suppliers to be licensed. The Minister of Economic Affairs and Climate Policy (the "**Minister**") grants a licence to any (aspiring) heat supplier that can satisfactorily prove that (i) he has the required organisational, financial and technical qualities for the proper performance of his duties and (ii) he may reasonably be deemed capable of fulfilling the obligations contained in Chap. 2 of the Dutch Heat Act.[25] There are provisions protecting vulnerable customers to be disconnected from the heat grid, for instance in the event of payment problems.[26] In the Act under consultation local communities will be attributed the power to designate local heat companies, giving the communities the ability to direct the heat transition in their regions in an efficient and sustainable way.

Pursuant to the operative Article 5 of the Dutch Heat Act, tariffs for the supply of heat are capped. The Authority for Consumers and Markets ("**ACM**") is charged with setting a maximum tariff. ACM determines the maximum tariff annually, basing its calculation on the no-more-than-otherwise principle by using the method of calculations as set out in the Heat Decision. This price-cap only applies to the supply of heat to heat consumers (who have a connection not exceeding 100 kilowatts). The price-cap imposed by ACM has been maintained in the revised version, despite criticism on the proper functioning of the price cap (ECORYS, 2016; Lavrijssen et al., 2013). The main concerns relate to the fact that the no-more-than-otherwise principle prevents cost-based pricing in the heat market, as well as the awareness that using this principle is unlikely to be future-proof in light of the increase in alternative ways of heating (ECORYS, 2016).

---

[25] Article 10 Dutch Heat Act.

[26] According to the Dutch Heat Act a vulnerable consumer is a consumer for whom the termination of the supply of heat would result in very serious health risks or for the consumer's household members.

The price ceiling was, and still is, based on what a heat consumer would have paid for the same amount of heat if he had used gas as energy source. On the basis of this no-more-than-otherwise principle, the maximum price is composed of two parts: (i) a usage-dependent part, expressed in euros per gigajoule and (ii) a usage-independent part, expressed in euros.[27] Every year, ACM recalculates the maximum tariff and publishes it in a decision. The published maximum tariff then applies until the 1st of January of the next year. Should ACM miss out on a year, the last applicable maximum tariff will continue to apply until the 1st of January of the year after the year in which ACM has again published a maximum tariff.[28] Once every two years, ACM collects, analyses and processes information concerning the development of returns made by heat suppliers. This is aimed at preventing excess profits made at the expense of heat consumers. ACM reports these findings to the Minister. The price of being connected to an existing heat network is also linked to the price of a connection to the gas grid. This is capped at the cost of being connected to the gas grid.[29]

With the energy transition leading to less use of gas and alternative heat sources on the rise, it will soon become untenable to apply the no-more-than-otherwise principle and to use gas prices as a standard. In order to promote competition between different heating technologies with an aim at reversing or stalling climate change, the idea of gas price as 'a standard price' has to be abandoned (Huygen et al., 2011). With different heating technologies becoming more prevalent and as the process of the phasing out of gas is on-going, gas will lose its dominant position as a heat source. A successful energy transition resulting in a sustainable energy system, benefits from rapid developments in heating technologies. From a perspective of fostering fair competition—on the merits of each heating technology—it will become untenable to consider the gas price as leading for heating. This is all the more so considering that the gas price has little to do with the costs of heat supplied via heat networks (Dutch Parliamentary Papers, 2016). In the draft Act on Collective Heat Systems the no-more-than-otherwise principle will gradually be phased out. The no-more-than-otherwise principle will be replaced by a regulatory methodology based on the calculation of the efficient costs of collective heat systems.

Recent Developments—Main Economic and Legal Challenges in the Dutch Heat Market

Regulation of the Dutch heat market, and in particular the Dutch Heat Act, is thus still largely based on old-fashioned assumptions that heat is provided by a central unit and that gas heating is the standard in the Netherlands. Developments in the Dutch heat market, however, are on the rise and show the need for a change in the organisation and regulation of the Dutch heat market as acknowledged by the

---

[27] Article 5(2) Dutch Heat Act. The Heat Decision provides a precise method of calculation.
[28] Article 5(3) Dutch Heat Act.
[29] Article 6 Dutch Heat Act.

Minister of Economic Affairs in the draft Act on Collective Heat Systems (Akerboom & Huygen, 2021). Several of these developments are discussed below.

Move Away from Gas

Save for exceptional circumstances, newly built houses are no longer connected to a gas grid since the 1st of July 2018.[30] The Dutch government further aims to have the use of natural gas (both for heating and cooking) phased out by 2050 (Ministerie van Economische Zaken en Klimaat, 2016). Contributing to the realisation of this development is the recent amendment to the Dutch Crisis and Recovery Act.[31] This amendment gives municipalities the possibility to cut-off existing households from the gas grid, as an experiment in anticipation of the revised Dutch Environment and Planning Act, that will enter into force in 2021 (Omgevingswet, 2019). As a consequence, district heating will cater for more households and play an important part of the Netherlands' plan to reduce carbon emissions (NRC Handelsblad, 2019).

The move away from gas exposes certain challenges that the current organisation of the heat market insufficiently addresses. For example, by capping the maximum price for heat consumers to the price of natural gas, the price for heat will depend on the price of heat paid by an ever-smaller group, that derives its heating from a non-preferred energy carrier. The heat transition also raises the important question of how vulnerable customers are protected in the heat transition and how this transition can be kept affordable for lower income groups.

Furthermore, whereas gas and electricity consumers are free to switch suppliers, similar possibilities do not yet exist for heat consumers due to technical and regulatory barriers. Hence, heat consumers are typically locked into long-term contracts whilst having no alternative heat source to switch to (Lavrijssen et al., 2013). In light of the foregoing, the move away from the use of gas also accentuates the importance of the question whether third-party access to heat networks should be implemented. For one, non-discriminatory network access may improve (the prospect of) competition on the heat market. This could attenuate the locked-in position that heat consumers find themselves in. Competition on a heat network could lead to lower prices or to the possibility of switching of supplier. In addition, the question is whether unbundling could play an efficient role in removing the incentive for vertically integrated firms to discriminate against (potential) competitors seeking network access.

Hence, the move away from gas stimulates changes and confronts the current regulatory set-up of the heat market with challenges that are presently unaccounted for. In the draft Act the minister does not envision separate social policy to protect vulnerable groups for possible negative consequences of the heat transition. For instance, it is not regulated how it can be prevented that the costs of disconnection from the gas grid will be passed on to an ever-smaller group of customers that stay

---

[30] Article 10(7)(a) and (b) Dutch Gas Act.

[31] Dutch Ministry of Economic Affairs and Climate Policy, Staatsblad (2019), 216.

connected to the gas grid. Furthermore, in the draft Act the Minister does not yet envisage an independent role for network companies, the introduction of unbundling requirements and regulated third party access, though these choices are still under debate. In order to reach optimal regulatory outcomes, advantages and disadvantages will have to be weighed and justifiable decisions have to be taken. The values of energy democracy and energy justice and the principles of good governance could prove helpful by providing a framework in which this decision-making can take place.

Climate Neutrality of Heat Networks

The increased reliance on heat as well as the transition to a low carbon economy push the need for new heat networks (Hoogervorst, 2017). In 2017, the PBL Netherlands Environmental Assessment Agency ("**PBL**")[32] took the view that the most economical way of supplying around 60–70% of the national demand for low temperature heat—would be through heat networks (Hoogervorst, 2017).

Existing heat networks are often fed from one large (fossil-fired) heat source or a limited number of heat sources from one owner (Hoogervorst, 2017). A low carbon economy requires, *i.a.,* heat networks that are fed from less-polluting or renewable sources. The generally smaller size of these heat sources will mean that heat networks may have to be fed from different heat sources, from different owners. This means that rules have to be in place for third-party access, to facilitate the use of heat from different sources and owners. In addition, a comprehensive set of rules also has to be in place in order to facilitate the integration and use of alternative heat sources for heating, like geothermal heat (Geothermie et al., 2018).

The PBL identified that currently, investment risks are high and financial returns low because the room for price increase is limited and there is an interdependence in actors required for success. Risks involved with investing in new heat networks need to be mitigated by reliable regulation. Furthermore, climate neutrality and energy efficiency are also linked to the use of the best placed heating solution given the circumstances. With regard to the choices that have to be made in order to stimulate a sustainable heat market, the values of energy democracy and energy justice and the principles of good governance can contribute by providing a framework within which the costs and benefits of different regulatory options can be weighed.

---

[32]PBL is the national institute for strategic policy analysis in the fields of the environment, nature and spatial planning in the Netherlands. PBL is part of the Ministry of Infrastructure, Public Works and Water Management.

Prosumers

With the number of prosumers in the energy sector increasing, it is clear by now that prosumers, also referred to as self-consumers or active consumers in EU regulation, are here to stay (Lavrijssen & Carrillo, 2017; Parag & Sovacool, 2016). The term prosumer generally refers to a consumer who generates (renewable) electricity for its own consumption, and who may store or sell self-generated electricity. The generation and/or storage of electricity is not the prosumer's primary commercial or professional activity (EUR-Lex 2018, 2019).[33]

Prosumers may also play a part in increasing the climate neutrality of the heat market (Brange et al., 2016). They typically generate energy from renewable sources. With an eye on the decarbonisation of the Dutch economy, this should only be encouraged and supported by regulation. The often-local character of a heat network may prove to be exceptionally compatible with small-scale prosumerism. In addition, prosumerism may respond to concerns expressed by way of the upcoming energy democracy and energy justice concepts in that it gives citizens a possibility to actively participate in the energy sector.

At the moment, unlike EU energy regulation, the Dutch regulation of the heat market and the new proposals of the Minister of Economic Affairs give little support to prosumers. Non-discriminatory network access, tailored to the needs of prosumers, may be needed. Prosumers may also experience hindrance from (vertically integrated) suppliers that are not interested in dealing with small-scale additions to their grid. Both unbundling and adequate supervision may under certain circumstances prove to be beneficial. Additionally, regulation and supervision of the heat market may also develop the rights and obligations of prosumers vis-à-vis customers and producers. Whether this is the case, will have to be analysed in light of the characteristics of different heat sources and networks. The principles of good governance are pertinent to carrying out the balancing exercise required to weigh the advantages and disadvantages associated with integrating prosumers into the heat market.

Digitalisation

Digitalisation of the energy sector involves the application of digital technology to the production, transport, distribution and supply of energy. Smart energy meters for example, enable consumers to monitor and manage their energy use and smart grids allow changes in supply and demand to be managed real-time by energy suppliers. The electricity sector is already benefiting from digitalisation, by providing consumers and suppliers alike with valuable insights upon which they can act (IEA, 2017).

---

[33] Article 2(14) Directive (EU) 2018/2001 of the European Parliament and of the Council of 11 December 2018 on the promotion of the use of energy from renewable sources (recast) and Article 2(8) Directive (EU) 2019/944.

Digitalisation efforts are also relevant for the heat market, especially in light of 4th and 5th generation heat networks (Lund et al., 2014). Increased digitalisation may lead to many possibilities, each with their own challenges. The use of smart meters for example, triggers questions on ownership, data-sharing and privacy. In addition, digitalisation necessarily requires investments in digital technologies, which lay bare the need for a stable investment environment. An environment that also provides rules applicable to the use of digital technology. Whereas digital technologies may provide more insights into demand and supply, there must be options to act upon these better insights—for example by switching to other energy sources or carriers (grid connection) and by regulating network access. Furthermore, the use and sharing of personal data should be regulated in line with the requirements of data protection and privacy regulation (EUR-Lex, 2016). Dutch regulation, however, does not address digitalisation of the Dutch heat market as yet. In order not to hamper (progress of the) digitalisation process, regulation has to facilitate the use of digital technologies in the heat network by protecting the rights of consumers and by empowering them to act on the possibilities available to them. The values of energy justice and energy democracy and the principles of good governance can provide guidance on how to identify the necessary regulatory changes and take the corresponding decisions.

Conclusion

It follows from the above that the (upcoming) developments in the Dutch heat market are interrelated, and together will lead to substantial reforms on the heat market. These developments, however, are not currently facilitated by a receptive regulatory environment. On the contrary, legislation of the Dutch heat market lags behind and is inhibiting these developments from taking full effect. Out-dated regulations governing the Dutch heat market in transition may stifle innovation and lead to a less-than-optimal regulatory environment (OECD, 1996). As the special position of vulnerable customers in the heat transition is not considered in the recent proposals for a new act, there is a risk that vulnerable customers are not sufficiently protected.

To let the heat market progress and contribute to a low-carbon economy, important decisions have to be made regarding the design of market regulation and supervision of the Dutch heat market. These decisions encompass, *i.a.*, (i) unbundling, (ii) third-party access, (iii) regulation and supervision—(iv) taking into account the rights and obligations of prosumers, consumers and vulnerable customers. To encourage adequate decision-making and soundly motivated decisions on these matters, energy democracy and energy justice and the principles of good governance ought to play a role in the weighing of the available options and their advantages and disadvantages. This may be done by way of the framework that the principles of good governance offer: a framework that promotes independence, accountability mechanisms, transparency requirements, possibilities for participation, effectiveness and efficiency.

Unbundling, third-party network access, adequate supervision and consumer/prosumer participation can significantly foster the development of the Dutch heat market. Taken together, addressing these challenges in a framework of good governance allows the Dutch heat market to develop and to contribute to a just energy transition. The following chapter will look into how these challenges have been overcome in the electricity sector. Citizen participation in the Danish heat market will be studied to draw inspiration regarding the way consumer participation can be enhanced.

The Dutch Electricity Market—Main Differences with Regulation of the Heat Market

The electricity market is subject to European legislation since the late 1990s. As of then, European regulation was enacted to remove national market structures in favour of internal electricity markets. EU regulation aimed to achieve a shift from national monopolies to fully integrated internal electricity markets. As a result, EU law requires the abolition of technical and regulatory barriers preventing the free flow of electricity across the EU (European Commission, 2015). Currently, a new shift is leading the EU from a fossil-fuelled electricity sector to one that is fuelled by renewable sources. In this shift, the opening up of the market to the possibility for new players to enter the market, has been of importance (Lindt, 2013). The use of renewable resources in the European electricity sector is given further shape via the 'Clean Energy Package'. This Clean Energy for all Europeans package comprises of several legislative initiatives aimed at the transition to a carbon-neutral economy (European Commission, 2019).[34] The last measures have been adopted in May 2019 and include new rules on the electricity market that, *i.a.*, will make it easier for renewable energy to be integrated into the electricity grid. The clean energy Package also promotes the empowerment of prosumers to let them benefit from the energy transition.

Unbundling

Requirements imposed by the first, second, and third Electricity Directive[35] (EUR-Lex, 1996, 2003, 2009) have led to effective unbundling in the electricity market in the Netherlands. The third Electricity Directive further strengthens the example set

---

[34] Relating to the energy performance in buildings, renewable energy, energy efficiency, governance regulation, electricity market design.

[35] The Third Electricity Directive (Directive 2009/72/EC) will be repealed with effect from 1 January 2021, by Directive (EU) 2019/944 of the European Parliament and of the Council of 5 June 2019 on common rules for the internal market for electricity and amending Directive 2012/27/EU (recast).

by the first and second Electricity Directives by imposing more stringent unbundling requirements on all players in the electricity market. Effective unbundling was aimed at and encompasses the separation of networks from activities of generation and supply of electricity[36] (EUR-Lex, 2009).

The third Electricity Directive offers three unbundling options regarding transmission systems: (i) ownership unbundling, where the undertaking that owns the transmission system acts as the transmission system operator[37] (EUR-Lex, 2009), (ii) an independent system operator, in case the transmission system belongs to a vertically integrated electricity company, a Member State may appoint an independent system operator that does not own a transmission system[38] (EUR-Lex, 2009), and (iii) an independent transmission operator, where the transmission system operator owning the transmission system is part of a vertically integrated electricity undertaking and has to adhere to the rules in chapter 5 of the Third Electricity Directive—to guarantee the independence of the transmission operator[39] (EUR-Lex, 2009). Distribution system operators that are part of vertically integrated firms are at least in terms of their legal form, organisation and decision making independent from other activities not relating to distribution[40] (EUR-Lex, 2009). Electricity undertakings continue to have to keep separate accounts for each of their transmission and distribution activities as they would be required to do if the activities in question were carried out by separate undertakings[41] (EUR-Lex, 2009).

The Netherlands took the unbundling requirements one step further by requiring that, ultimately, the shares in the distribution system operators in the electricity sector and gas sector have to be held by the local and regional public authorities. The system operators cannot belong to a group of undertakings active in the production, trade and supply of energy (ownership unbundling).[42] Hence, far-reaching unbundling requirements are imposed upon players in the Dutch electricity market. However, similar requirements are not enacted for the Dutch heat market. When it comes to ownership, a telling difference is that heat networks may be owned by private firms, whereas energy networks may not.[43]

---

[36]Recital 9 Third Electricity Directive.

[37]Article 9(1)(a) Third Electricity Directive.

[38]Article 9(8)(a) and Article 13 Third Electricity Directive.

[39]Article 9(8)(b) and Article 13 Third Electricity Directive.

[40]Article 26 Third Electricity Directive.

[41]Article 31 Third Electricity Directive.

[42]Article 98 Dutch Electricity Act 1998. See also HR 26 June 2015, ECLI:NL:HR:2015:1727 (Dutch State v. Essent). The Dutch Supreme Court thereby complied with the preliminary ruling of the Court of Justice of 22 October 2013, case C-105/12 – C-107/12, ECLI:EU:C:2013:677.

[43]Article 93a Dutch Electricity Act 1998.

Third-Party Network Access—Electricity

Third-party network access in the electricity market is guaranteed by the third Electricity Directive and by the recast Electricity Directive, stipulating that Member States need to implement a system of third-party access to the transmission and distribution systems based on published tariffs, applicable to all eligible customers. The system for network access has to be applied objectively and without discrimination between system users. The (calculation of the) access tariffs is subject to prior approval, promoting the possibility for effective access[44] (EUR-Lex, 2009, 2019).

Third-party access systems are established both in the Dutch electricity market and Dutch gas market but currently not in the Dutch heat market. In the explanatory memorandum to the revised Dutch Heat Act, the Dutch legislator indicated that it will not impose measures of regulated third-party network access. This choice is influenced by the physical characteristics of the 'heat product', the large differences between heat networks and the current state of market development (Dutch Parliamentary Papers, 2016). Instead, the Dutch Heat Act foresees in a negotiated access requirement.[45] This provision applies to the network operator and the supplier(s) using the network.

Supervision

The third Electricity Directive stipulates that each Member State has a designated national regulatory authority in charge of supervising the electricity market[46] (EUR-Lex, 2009, 2019). This national regulatory authority is legally and functionally independent from other entities. Its staff acts independently from market interests and cannot take direct instructions from any government or entity. Decision-making is autonomous and the national regulatory authority has a separate allocated budget. In addition, its board or top management is appointed for a fixed term.

In the Netherlands, ACM has been entrusted with supervising the electricity market. ACM is independent to the extent that it complies with the relevant European requirements that impose a level of independence and has a certain degree of flexibility in how to use its powers to adopt decisions. However, it should be noted that there may be a tension between the independence and the flexible powers that ACM has vis-à-vis the requirement that it has to be held accountable for its actions (Lavrijssen & Ottow, 2012; Lavrijssen, 2019). Such tensions are tangible and need constant and careful consideration to guarantee a well-functioning independent supervisory authority in the electricity sector.

---

[44] Article 32 Third Electricity Directive. Ártice 6, Directive (EU) 2019/944 of the European Parliament and of the Council of 5 June 2019 on common rules for the internal market for electricity and amending Directive 2012/27/EU (Text with EEA relevance.), OJ L 158, 14.6.2019, p. 125–199.
[45] Article 21 Dutch Heat Act.
[46] Article 35 Third Electricity Directive and Article 57 of Directive 2019/944.

Assessment

Whereas the electricity market is liberalised by way of unbundling, third-party network access and supervision that are regulated on an EU level, the same does not apply to all parts of the energy sector. The opening up of the electricity market has paved the way for the energy transition to take effect in the electricity sector by allowing renewable sources and alternative ways of generating electricity to prosper. Most notably, the Clean Energy Package has taken up the role of leading the electricity market into a low-carbon future and consumer/prosumer empowerment. Hence, unbundling, third-party network access and supervision are not the only factors encouraging the energy transition in the electricity sector, but they certainly are the first steps in the right direction. In this light, an assessment of whether the Dutch heat market is ready for the changes brought on by the energy transition is called for.

The most relevant principles of good governance, the sector-specific concepts of energy justice and energy democracy, as well as the main recent developments have been set out above. The assessment in this chapter, will evaluate the four main challenges—unbundling, third-party access, supervision and consumer/prosumer participation—in light of the framework provided by the principles of good governance and the concepts of energy democracy and energy justice. Considering the different stages of technological developments and the different characteristics of the networks, different solutions may be appropriate. For a well-functioning heat market, it is essential that appropriate regulation is in place. The principles of good governance can play a role in reaching balanced decisions.

Unbundling

The following levels in the district heating chain can be distinguished: (i) production, (ii) transport/distribution and (iii) delivery of heat to heat consumers. This chain is commonly organised in the Netherlands by way of an integrated model; meaning that the production of heat, the ownership of the heat network and the supply of heat are all combined in one company (ECORYS, 2016).

Keeping in mind the current characteristics of the heat market as set out above, several studies have found that unbundling is not yet feasible for the existing heat networks (ECORYS, 2016; SiRM, 2019; Tieben & Van Menno, 2018). Recommendations against unbundling are mainly based on research into 3rd generation heat networks—the most common type in the Netherlands. The recommendation against unbundling holds true for all its forms: accounting, organisational, legal, and in terms of ownership (Tieben & Van Benthem, 2018). In particular, technical limitations of most functional heat networks prevent unbundling from remedying the existence of natural monopolies. Hence, at the current stage of economic and technological development, unbundling would lead to disproportionate costs. These costs relate to the separation of the different levels of the chain from production to delivery to heat consumers—without (as yet) adding any value to heat consumers (ECORYS, 2016).

Unbundling Assessed

In spite of the fact that unbundling currently is recommended against, unbundling can bring significant benefits under the right circumstances. For example, unbundling takes away the incentive for vertically integrated firms to favour their own (up- and downstream) activities and decreases the risk of overcharging consumers or providing suboptimal service. Hence, it takes away the incentive for dominant firms to abuse their position, ranging from margin squeeze to overcharging or providing suboptimal service to locked-in heat consumers. To some extent, unbundling may also take away cross-subsidisation and prevent unfair competitive advantages to inefficient (parts of) undertakings (Steyn, 2014). Moreover, the development of technologically advanced heat networks is on the way and the importance of heat markets is on the rise. This combination of factors calls for a future-proof organisation of the Dutch heat market.

What is important to keep in mind, is that the Dutch heat market consists of a large variety of heat networks. As has been laid out in chapter three, differences in heat networks range from type, size, ownership structure and heat source to the age of heat networks. In light of the many differences it may be advisable to have unbundling requirements in place for newer generation, larger scale (fourth of fifth generation) heat networks where the advantages of unbundling could outweigh the implementing costs. This would provide for the opportunity to prevent abuse of dominance issues by way of imposing unbundling requirements that pre-empt possible abusive behaviours facilitated by vertical integration. This is a strong indication that customised regulatory solutions are called for as a one-size-fits-all approach would not be in line with the structure of the heat market.

Independence between the unbundled firms—especially when it concerns legal or ownership unbundling—can also bring benefits in the field of transparency. Cost-transparency may be increased by making it mandatory for firms to split activities and isolate costs to keep them where they are incurred. This means that cross-subsidising risks diminish, or at the very least, become visible. That may bring on an additional incentive for cost reduction and the promotion of efficiency, especially in the case of ownership unbundling where all firms active on the heat market must survive on their own merits.

Customised regulatory solutions regarding unbundling—*i.e.*, having different degrees of unbundling requirements in place according to the type of heat network—also relate to the principle of effectiveness and provide for a proportionate course of action. The imposition of unbundling requirements to older heat networks would not be efficient, as has been concluded before (ECORYS, 2016; Tieben & Van Benthem, 2018). However, not to impose unbundling requirements at all—as is currently the case—is likewise not effective, as some heat networks benefit from unbundling. This means that there is a very strong case to make for customised solutions within the framework of the organisation of the Dutch heat market. It is noteworthy that such a solution may also be favourable from the viewpoint of efficiency. Unbundling can spur the efficiency of firms. In order to reach the most efficient situation for each heat network, unbundling requirements could range from accounting, organisational, legal, and ownership unbundling. The principles of effectiveness, efficiency and transparency can guide the choices to be made when the costs and benefits of different options are assessed.

Unbundling Conclusion

The prevalence of vertically integrated companies in the heat market added to the fact that the heat market is comprised of (necessarily local and natural) monopolies, leaves heat consumers in a vulnerable position. Depending on the economic and technological circumstances of the heat network at stake, heat consumers could benefit from the advantages that unbundling typically brings. While currently the absence of unbundling requirements is explicable in light of the technical limitations of the majority of heat networks in use in the Netherlands, the introduction of technological advancements could mean that—in the absence of unbundling requirements—regulation of the heat market has to catch-up with the market and new possibilities for adding more (lower temperature) heat sources to the heat network. In light of the fact that the diversification in heat networks will increase as new heat networks are being built, one-size-fits-all solutions cannot lead to satisfactory results. Regulation should not limit technological choices; it should provide for requirements based on objective criteria according to the type of heat network (third, fourth of fifth generation networks). This ensures proportionate regulation that provides for the possibility of imposing varying degrees of unbundling requirements and does not hinder technological developments that can engage energy consumers/prosumers more actively in the heat market (ECORYS, SEO, 2020).

Third-Party Access

The revised Dutch Heat Act, that has entered into force on the 1st of July 2019, provides for a light version of negotiated access and not for a regime of regulated third party access like in the Electricity Act 1998. The act's new Chap. 6 obliges network operators (and heat suppliers active on the network) to respond to requests from heat producers and consult with them on access to the network for the purpose of transporting heat. Both the network operator and the heat supplier have to disclose information to the requesting heat producer regarding, *i.a.*, the available transmission capacity, heat demand and production capacity.[47]

---

[47] Article 21 Dutch Heat Act.

Considering the technological limitations of most of the heat networks in the Dutch heat market, third-party access is still—but conceivably not for too long—something for the future. This is mainly due to the fact that heat networks are closed loops. Older generations of heat networks distribute hot water of around 90–95 °C or steam and are therefore highly sensitive to change. Change in the form of adding extra parties to the loop may alter the temperature of the water. As a consequence, third-party access may not be a proportionate and efficient solution because the costs of implementing and enforcing it would outweigh its benefits.

Third-Party Access Assessed

Technology is advancing and 4th and 5th generation district heating (and cooling) networks will be up and coming (Buffa et al., 2019). 5th generation heat networks (sometimes also referred to as heat sharing networks) distribute water at a close to ambient ground temperature. Thereby both heating and cooling is facilitated: by means of a heat pump heat is either delivered to a household, or excess heat is fed into the network. This means that transport losses are significantly limited and installing costs are lowered. Because the water temperature is relatively low, it may also become easier to add parties onto the network.

With the perspective of the principle of independence in mind, the Dutch Heat Act's provision on negotiated network access—with no dispute resolution mechanism in place—might better be replaced by non-discriminatory access rights in the case of technologically more advanced heat networks. Enforceable access rights, aimed at network access on fair terms, protect parties seeking access to the network as it makes it difficult for the network owner, to refuse to grant access on invalid grounds. At the same time, in case third-party access is refused on allegedly invalid grounds, the party seeking access is in a better position to enforce its rights, compared to when only an obligation to negotiate would exist. Independence also means that the owner of the heat network is either independent from all the parties on the network (which would be the case for heat networks where ownership unbundling is required) or the owner is prevented from treating certain parties on the network more favourably than others.

In order to curb the dominant position of the network owner, who is in charge of granting third-parties access to the network, a fair access mechanism should be in place. It is likely that regulated third-party access increases transparency to the benefit of heat consumers and other stakeholders. Parties seeking network access, will have to be given insight into the access costs and could compare access prices from different heat networks that qualify for third-party access regulation. While it is essential to keep in mind that the many differences between types of heat networks mean that direct comparisons between access prices may not be drawn, a cost breakdown could still be helpful in assessing whether a fair price is imposed. Independent regulation could indicate the composition of access prices by cost breakdown elements. This may also increase cost-efficiencies.

Heat consumers having the means to generate their own heat (prosumers) should have the possibility—like in the electricity sector—to be active on the wholesale heat markets via heat networks that can technically (and economically) support this[48] (EUR-Lex, 2019). This will mostly apply to future 4th and 5th generation heat networks. Third-party access rights specifically aimed at prosumers can facilitate this. Such rights are valuable from the perspective of energy justice and energy democracy, as they provide heat consumers with a chance to be actively involved in and shape the heat market.

Third-Party Access Conclusion

In light of the possibilities offered by 5th generation heat networks, as well as technological advancement in general, third-party access may thus prove to be beneficial to foster competition on the heat market and consumer choice. However, as not all heat networks are technologically similar, customised solutions could be called for. For example, different categories of heat networks could be identified. Whether third-party access rights can then be given to parties, would depend on the type of heat network and the heat sources available. For this assessment, the values of energy justice and energy democracy and the principles of good governance can provide a framework that safeguards justified decision-making. In providing this type of 'flexibility', a consistent application of the principles of good governance is needed to safeguard that regulation—and regulatory choices—do not become inconsistent. This requires the regulatory authority to have discretionary powers, so that it can assess which heat network should be subject to what kind of regulation. However, more discretionary powers for a regulatory authority increase the tension with accountability requirements. Hence, accountability mechanisms become more important and should carefully be monitored. The following section will look into the current supervisory set-up of the Dutch heat market.

Supervision and Participation

ACM is the designated regulatory authority in charge of implementing and enforcing the Dutch Heat Act.[49] This means that ACM is in charge of setting the maximum heat price.[50] As explained, the maximum heat price is linked to the price of natural gas currently, but this system will be replaced by a cost based method for the calculation

---

[48] See Article 15 of Directive (EU) 2019/944 that stipulates that "*Member States shall ensure that final customers are entitled to act as active customers without being subject to disproportionate or discriminatory technical requirements, administrative requirements, procedures and charges, and to network charges that are not cost-reflective.*".

[49] Article 14 Dutch Heat Act.

[50] Article 5 Dutch Heat Act.

of the heat tariffs. Once every two years, ACM also monitors the development of the return rates in the heat supply market.[51] In addition, ACM is authorised to carry out measurements (or let them be taken) at heat producers, heat suppliers, and heat consumers.[52] Should ACM find that the Dutch Heat Act is not complied with, it may impose binding codes of conduct, periodic penalty payments and administrative fines.[53]

Supervision Assessed

If the organisation of the Dutch heat market is overhauled to support a more holistic energy sector, the current regulatory framework and supervisory set-up will no longer do. Supervision arrangements should carefully be coordinated to match the future of the heat market.

Customised solutions regarding unbundling and third-party access—as well as the modernisation of the heat market—require supervision that does not hinder, but rather stimulate the development of the heat market without 'prescribing' or favouring certain technological outcomes. This implies that ACM ought to be independent from market parties—as regulatory capture may result in favouring a certain technology over another—but also, to some extent, from political influence. This is already partially guaranteed because ACM employs specialists that are expected to have the best interests of the heat market at heart, rather than short-term political gains. To maintain this, it is preferable that ACM stay away from politics on a daily basis. Politicians should decide on the long-term goals, to which ACM in its capacity as the regulatory authority of the heat market should tend by decision-making on a day-to-day basis.

When it comes to discretionary powers, ACM is not currently adequately equipped to impose and enforce different regulatory requirements for different heat networks as the Dutch Heat Act does not provide a basis for a customised regulatory regime. Nevertheless, as a consequence of the great variety between the characteristics of established and future heat networks, it is important that the regulatory framework of the heat sector supports proportionate regulation and tailormade solutions. Effective regulatory oversight requires ACM to have adequate powers to facilitate the integration of competitive clean heat carriers in the energy system. Having the possibility to impose and specify unbundling and third-party access requirements in the heat sector—and to finetune them depending on the type of network— is a step in the right direction. Such levels of discretionary powers provide for flexibility in the actions of ACM. Increased flexibility though, raises questions on whether the current accountability mechanism will be sufficient to address any concerns stemming from an increase in flexibility on the side of ACM.

---

[51] Article 7 Dutch Heat Act.

[52] Article 16 Dutch Heat Act.

[53] Article 17 and 18 Dutch Heat Act.

ACM is part of the ministry of Economic Affairs and Climate Policy, and as such, 'inevitably' renders account to the Minister. In its turn, the Minister renders account to the Dutch government. Separately, ACM also publishes an annual report in which it reports the work it has performed the previous year. Whereas stakeholders have few possibilities to hold ACM accountable on the basis of the annual report, it does serve a valuable function in terms of transparency. While ACM already acts in a transparent way to a great extent—for example, by publishing its decisions and issuing guidelines—this needs to be maintained, and possibly even upped. A duty of transparency contributes to fairness in decision-making and provides stakeholders with the possibility to keep ACM in check. This is an important element of accountability.

Consumer/Prosumer Participation

The accountability element can also be linked with the importance of stakeholder participation. With the transition towards a low carbon economy on the way, a broad support base is needed to reach decarbonisation goals. Energy justice and energy democracy require strong participation possibilities that can shape the support given to the energy sector in transition. Heat consumers participation in the heat market can provide such support. As it is likely that most heat consumer will remain locked in for a relatively long time in natural monopolies facilitated by older generation heat networks (and the lack of alternative heating), heat consumers could get involved with heat producers, suppliers and/or network owners (depending on whether or not they are vertically integrated). This involvement would be a form of substantive participation and can take various forms. For example, heat consumers could become 'shareholders' in already existing arrangements, or set up a new communal heat network in which they are directly involved regarding ownership, decision-making etc. This way, heat consumers have direct influence on their heat supply arrangement by way of profit sharing and decision-making—thereby potentially increasing the chances that older generation heat networks are being made more sustainable. Consumer engagement may spur environmentally friendly solutions by way of pressure and direct involvement. Benefits of consumer involvement may especially apply to newer generation heat networks, where sustainability can play a larger role right from the start by facilitating acting together, the connection of different grids in the energy sector, prosumerism, renewable energy sources, and the interlinking of various local energy-efficient ideas (Mendonça et al., 2009; Ropenus & Henrik, 2015). Despite its promising benefits, consumer participation, as well as the role of prosumers, receives little attention in the Dutch heat market and in the draft Act on Collective Heat Systems. This is different in Denmark, where consumer participation is part and parcel of the heat market. Because the Netherlands may benefit from consumer participation in the heat market, Danish experiences with consumer participation will be explored further below for inspiration.

Supervision and Participation Conclusion

Whereas the current form of supervision in the heat market caters to the principles of independence, accountability and transparency, there is little room left for supervision to adjust to (economic and technological) developments. The developments described in paragraph four are testament to the fact that a more flexible regulatory approach is necessary to lead the heat market into the future and to facilitate the developments related to fifth generation heat networks and prosumerism. There is a need for differentiating between types of networks and heat sources to offer tailor made solutions. And for tailor made solutions to be effective, it has to be made sure that costs and benefits can be weighed in line with energy justice and energy democracy and the principles of good governance. The economic and technological developments taking place throughout the energy sector in transition, also require a more holistic supervisory model. Different parts of the energy market—gas, electricity, heat, etc.—will have to become more of a whole in order to reach optimal and just results in light of the energy transition. This means that there is a need for a flexible regulatory model that can adjust to technical and economic developments. The regulatory model should also provide for consumer/prosumer participation possibilities to create support and involvement and to ensure that the energy transition can take place in a just way. The below paragraph will briefly explore how participation plays a role in Denmark's heat market.

Consumer Participation: Denmark

In terms of citizen participation, Denmark's heat market is one that can serve as an example. Not only has the extensive use of heat networks in Denmark a long history, Denmark also leads the way when it comes to citizen-involvement in the heat market (Huygen et al., 2019).

The heat market in Denmark can be laid out in a succinct manner as follows. The Danish Utility Regulator[54] is the independent regulatory authority[55] overseeing the Danish heat market and is in charge of applying the Danish Heat Supply Act (Danish Energy Agency, 2017). The Danish Heat Supply Act promotes the most economic and environmentally friendly use of energy for heat in order to reduce Denmark's dependence on fossil fuels.[56] This Act stipulates that municipalities are in charge of providing for collective heat supply, whereby stakeholders are involved in the preparation of municipal heat supply plans.[57] Once heat supply is up and running, the Danish Utility Regulator is tasked with, *i.a.*, monitoring and supervising the

---

[54]Until 1 July 2018 the Danish Energy Regulatory Authority (DERA).

[55]See chapter 2 of the Act establishing the Danish Regulatory Authority. https://www.retsinformat ion.dk/eli/ft/201712L00164.

[56]Chapter 1 §1 of the Danish Heat Supply Act.

[57]Chapter 2 §3 of the Danish Heat Supply Act.

prices charged by the heat supplier. Consumer prices are cost-based, and the Danish Heat Act specifies which costs and expenses may be included in the heat price.[58] To allow the Danish Utility Regulator to monitor the prices, suppliers have to notify the prices and conditions they apply.[59] In addition to prescribing a cost-based pricing approach, the Danish Heat Act allows for the imposition of maximum prices for heat generated by waste incineration.[60] The rationale for this option is to safeguard that heat consumers only pay for the costs attributable to the production of heat.

With regard to consumer participation, it should be noted that the Danish Heat Supply Act grants significant consumer rights. Communal ownership—or, in the absence thereof, substantive heat consumer-participation, is preferred. For example, before a heat supply plant is sold to a buyer other than a municipality, the consumers who are purchasing heat from that plant, must be offered the possibility to jointly acquire the plant at market price.[61] All plants that are not owned by municipalities and/or the plants' heat consumers have to be managed by an independent company.[62] Furthermore, when consumers do own a heat supply plant, the majority of the company's members of the board of directors must be elected by the consumers whose properties are connected to the company's plant or by one or more municipal boards in the company's supply area.[63] In the absence hereof, the Danish Heat Supply Act foresees in the establishment of a mandatory consumer representative board consisting of 11 members. These members must be elected by the consumers whose properties are connected to the heat supply plant. The consumer representative board on its turn, elects the majority of the board members of the heat supplying company.[64]

In Denmark, consumers thus directly participate at company level—either directly via ownership or indirectly via representation (Chittum & Østergaard, 2014). This has led to a cost-effective heat market with locally empowered heat consumers. As Danish heat consumers are "*confident that their needs are being adequately and accurately represented to company decision-makers*", increased consumer participation may also be beneficial to Dutch heat consumers (Chittum & Østergaard, 2014).

---

[58]Chapter 4 of the Danish Heat Supply Act.

[59]Chapter 4 §23b of the Danish Heat Supply Act.

[60]The Danish Utility Regulator publishes its decision on the maximum price for heat generated by waste incineration online.

[61]Chapter 4 §23f of the Danish Heat Supply Act.

[62]Chapter 4 §23g of the Danish Heat Supply Act.

[63]Chapter 4 §23h of the Danish Heat Supply Act.

[64]Chapter 4 §23i of the Danish Heat Supply Act.

Conclusion

This article observed the trend that the values of energy justice and energy democracy are increasingly colouring the principles of good market regulation and supervision and deserve more attention in the regulation of the Dutch heat transition. The heat transition in the Netherlands significantly impacts the position of consumers, prosumers and vulnerable customers, as an ever-larger group of consumers will be disconnected from the gas grid and will be connected to heat networks. Energy democracy and energy justice are important values that should guide policy-makers in making choices that affect consumer participation and the protection of vulnerable customers. In this regard more (flexible) regulations are needed to ensure that consumer and prosumer participation is promoted and safeguards are put in place to protect vulnerable customers.

Increased consumer participation is important in light of achieving energy justice and energy democracy—energy specific concepts that are gaining influence when interpreting and applying the principles of good governance. Both are based on the awareness that the energy transition is a matter for all citizens of the European Union (and world-wide). Denmark has extensive experience with substantive consumer participation -providing for consumer ownership and significant opportunities for consumer representation in the boardroom of local heat companies- and could thus serve as an example for the Netherlands to take inspiration from. Because it is likely that most heat consumers will remain locked in for a relatively long time in natural monopolies facilitated by older generation heat networks and given the lack of alternative heating options, substantive consumer-participation could yield positive results regarding community engagement in heat network management and heat supply.

Considering that the Dutch heat market has still a lot to gain in terms of flexible regulation and supervision as well as participation, the principles of good market regulation and supervision have a significant role to play in modernising the regulation of the Dutch heat market. Against the background of the values of energy democracy and energy justice, the principles of good regulation can provide a framework within which advantages and disadvantages can be weighed of regulatory choices to be made by the Minister of Economic Affairs and the public authorities that will be involved in applying the regulatory framework (local authorities and the ACM). The values of energy democracy and energy justice and the principles of good regulation can also be embedded in the law by the legislator to ensure the accountability of the relevant public authorities and the independent regulatory authority vis-à-vis the consumers/prosumers when regulating the heat market.

The Minister of Economic Affairs has opened a public consultation on the draft Act on Collective Heat Systems. Though this draft act is still under debate, initially it provided little flexibility for finetuning regulatory requirements to the specific technological and economic characteristics of different types of heat systems. Local authorities will be responsible for designating integrated heat companies based on the integrated market model, with no independent network operator and no open network access regime. There will be a special regime for large scale regional networks, that will have to be unbundled from the production and the supply of heat and that are governed by regulated third party access. The lack of flexibility could play up at not providing unbundling and third-party access when 4th and 5th generation heat networks become in use, as these types of networks could technologically be ready for competition on the heat market and could facilitate the integration of prosumers. As the electricity market provides clear evidence that unbundling, third-party access and prosumerism have many advantages, it would be constructive to benefit from similar advantages in the heat market. A larger need for flexibility is then justified because of the differences between types of heat networks. It would allow the local authorities together with ACM to assess on a case-by-case basis whether unbundling and/or third-party access would be beneficial for consumers and prosumers and preferable for specific heat networks and enabling them to specify regulatory requirements.

Regulation will have to be up to date with the energy transition that relies on and stimulates new forms of energy generated with sustainable sources and new ways of storing energy. This means that a more holistic regulatory approach is needed connecting different energy networks and different energy carriers. The development of a new Act on Collective Heat Systems provides for the perfect opportunity to let flexibility and consumer participation enter the Dutch regulatory set-up of the heat market to ensure it is well-equipped for the future. This way regulation can facilitate a just heat transition in which consumer access to affordable, sustainable and secure heat supply is safeguarded and where there are opportunities for all consumers to participate in the heat market in a meaningful way. This opportunity should not be missed.

References

ACM. (2018). *Besluit tot vaststelling van maximumprijs en de berekening van de eenmalige aansluit-bijdrage en het meettarief warmteverbruik per 1 januari 2019.* https://www.acm.nl/sites/default/files/documents/2018-12/besluit-maximumprijs-warmte-2019.pdf. Accessed January 7, 2021.

Addink, H. (2005). Principles of good governance: Lessons from Administrative Law. In D. Curtin, & R. Wessel (Eds.), *Good governance and the European Union—reflections on concepts, institutions and substance* (pp. 21–47). Intersentia.

Aelen, M. (2014). *Beginselen van goed markttoezicht—Gedefinieerd, verklaard en uitgewerkt voor het toezicht op de financiële markten.* Boom Juridische uitgevers.

Akerboom, S., & Huygen, A. (2021). Regulatory aspects of the urban heat transition: The Dutch heat act. *Carbon and Climate Law Review, 14*(4), 281–293. https://doi.org/10.21552/cclr/2020/4/7.

Alarcón, Cristian, & Chartier, Constanza. (2018). Degrowth, energy democracy, technology and social-ecological relations: Discussing a localised energy system in Vaxjö, Sweden. *Cleaner Production, 197*(2), 1754–1765.

Alemanno, A. (2013). Unpacking the principle of openness in EU law—transparency. *Participation and Democracy. European Law Review, 39*(1), 72–90.

Andrews, M. (2008). *Good government means different things in different countries.* HKS Working Paper No. RWP08-068. https://www.hks.harvard.edu/publications/good-government-means-different-things-different-countries. Accessed January 8, 2021.

Baldwin, R., Martin, C., & Martin, L. (2012). *Understanding regulation theory, strategy, and practice*. Oxford University Press.

Bickerstaff, k, Gordon, w, & Harriet, Bulkeley (Eds.). (2013). *Energy justice in a changing climate: Social equity and low-carbon energy*. Zed Books.

Boersma, M. A. M. (2015). *Nutsbedrijven: Quo Vadis?* https://www.tias.edu/docs/default-source/Kennisartikelen/150312_oratie_prof-boersma.pdf?sfvrsn=0. Accessed January 7, 2021.

Bovens, M. (2006). Analysing and assessing public accountability. A conceptual framework. *European Law Journal, 13*(4), 447–468.

Brange, L., Englund, J., & Lauenburg, Patrick. (2016). Prosumers in district heating networks—a Swedish case study'. *Applied Energy, 164*, 492–500.

Bruijn, H., & Dicke, W. (2006). Strategies for safeguarding public values in liberalized utility sectors. *Public Administration, 84*(3), 717–735.

Buffa, S., Cozzini, M., D'Antoni, M., Baretieri, M., & Fedrizzi, R. (2019). 5th generation district heating and cooling systems: A review of existing cases in Europe. *Renewable and Sustainable Energy Reviews, 104*, 504–522.

Bukento, A. (2016). Sharing energy: Dealing with regulatory disconnection in dutch energy Law. *European Journal of Risk Regulation, 4*, 701–716.

CBS (2019). Nederland langs de Europese meetlat in *Herniewbare energie*. https://longreads.cbs.nl/europese-meetlat-2019/hernieuwbare-energie/. Accessed January 5, 2021.

CBS and ECN. (2017). *Monitoring warmte 2015*. https://www.cbs.nl/nl-nl/publicatie/2017/15/monitoring-warmte-2015. Accessed January 19, 2021.

Chittum, A., & Østergaard, P. (2014). How Danish communal heat planning empowers municipalities and benefits individual consumers. *Energy Policy, 74*, 456–474.

Council of Europe. (2008). *12 Principles of good governance*. https://www.coe.int/en/web/good-governance/12-principles. Accessed January 15, 2021.

Danish Energy Agency. (2017). *Regulation and planning of district heating in Denmark*. https://ens.dk/sites/ens.dk/files/Globalcooperation/regulation_and_planning_of_district_heating_in_denmark.pdf. Accessed January 8, 2021.

Danish Energy Agency. (2017). *Regulation and planning of district heating in Denmark*. https://ens.dk/sites/ens.dk/files/Globalcooperation/regulation_and_planning_of_district_heating_in_denmark.pdf. Accessed January 25, 2021.

De Visser, M. (2009). *Network-based governance in EC law: The example of EC competition and EC communications law*. Hart Publishing.

Den Hertog, J. (2010). Review of economic theories of regulation. *Discussion Paper Series/Tjalling C. Koopmans Research Institute, 10*(18), 1–59.

Dutch Parliamentary Papers 2002–2003, 29048 Nr. 3. (2003). Available online: https://zoek.officielebekendmakingen.nl/kst-29048-3.html. Accessed January 21, 2021.

Dutch Parliamentary Papers 2016–2017, 34723. (2017). Nr. 3. https://zoek.officielebekendmakingen.nl/kst-34723-3.html. Accessed January 21, 2021.

Dutch Parliamentary Papers 2016–2017, 34723. (2017). Nr. 6. https://zoek.officielebekendmakingen.nl/kst-34723-6.html. Accessed January 21, 2021.

ECORYS. (2016). *Evaluatie Warmtewet en toekomstig marktontwerp warmte*. http://www.nietmeerdan.nl/downloads2/20160209%20evaluatie%20ecorys%20eindrapport%20def.pdf. Accessed January 7, 2021.

ECORYS, ECN. (2014). *The Role of DSOs in a Smart Grid environment*. https://ec.europa.eu/energy/sites/ener/files/documents/20140423_dso_smartgrid.pdf. Accessed January 7, 2021.

ECORYS, SEO. (2020). *Regulering van de Nederlandse warmtevoorziening, analysekader en beleidsadvies in opdracht van Netbeheer Nederland. Report 2020–23*. https://www.seo.nl/publicaties/regulering-van-de-nederlandse-warmtevoorziening/. Accessed January 24, 2021.

Edens, M. (2017). Public value tensions for Dutch DSOs in times of energy transition: A legal approach. *Competition and Regulation in Network Industries, 18*(1–2), 132–149.

Eerste Kamer der Staten-Generaal. (2003). *Initiatiefvoorstel-Ten Hoopen en Samsom Warmtewet.* https://www.eerstekamer.nl/wetsvoorstel/29048_initiatiefvoorstel_ten. Accessed January 21, 2021.

EUR-Lex. (1991a). *Judgment C-18/88 of the European Court of Justice of 13 December 1991, Régie des télégraphes et des téléphones v GB-Inno-BM SA, ECLI:EU:C:1991:474.* https://eur-lex.europa.eu/legal-content/EN/TXT/?uri=CELEX%3A61988CJ0018. Paragraph 14. Accessed January 15, 2021.

EUR-Lex. (1991b). *Judgment C-202/88 of the European Court of Justice of 19 March 1991, French Republic v Commission of the European Communities, ECLI:EU:C:1991:120.* https://eur-lex.eur opa.eu/legal-content/EN/TXT/?uri=CELEX%3A61988CJ0202. Paragraph 51. Accessed January 15, 2021.

EUR-Lex. (1993). *Judgment C-69/91 of the European Court of Justice of 27 October 1993, Decoster, ECLI:EU:C:1993:853.* https://eur-lex.europa.eu/legalcontent/EN/TXT/?uri=CELEX%3A6199 1CJ0069. Paragraphs 17, 19 and 22. Accessed January 15, 2021.

EUR-Lex. (1996). *Directive 96/92/EC of the European Parliament and of the Council concerning common rules for the internal market in electricity (First Electricity Directive).* https://eur-lex.europa.eu/legal-content/EN/TXT/?uri=CELEX%3A31996L0092. Accessed January 23, 2021.

EUR-Lex. (2003). *Directive 2003/54/EC of the European Parliament and of the Council concerning common rules for the internal market in electricity and repealing Directive 96/92/EC (Second Electricity Directive).* https://eur-lex.europa.eu/legal-content/EN/TXT/?uri=celex%3A3 2003L0054. Accessed January 23, 2021.

EUR-Lex. (2009). *Directive 2009/28/EC of the European Parliament and of the Council on the promotion of the use of energy from renewable sources and amending and subsequently repealing Directives 2001/77/EC and 2003/30/EC.* https://eur-lex.europa.eu/legal-content/EN/ALL/?uri= CELEX%3A32009L0028. Paragraphs 16–62. Accessed January 17, 2021.

EUR-Lex. (2009). *Directive 2009/72/EC of the European Parliament and of the Council concerning common rules for the internal market in electricity and repealing Directive 2003/54/EC (Third Electricity Directive).* https://eur-lex.europa.eu/legal-content/EN/ALL/?uri=celex%3A3 2009L0072. Accessed January 23, 2021.

EUR-Lex. (2010). *Judgment C-246/09 of the European Court of Justice of 8 July 2010, Susanne Bulicke tegen Deutsche Büro Service GmbH, ECLI: EU: C: 2010: 418.* https://eur-lex.europa.eu/legal-content/EN/TXT/?uri=CELEX%3A62009CJ0246. Paragraph 25. Accessed January 17, 2021.

EUR-Lex. (2013). *Judgment C-280/11 P of the European Court of Justice of 17 October 2013, Council of the European Union v Access Info Europe, ECLI:EU:C:2013:671.* https://eur-lex.eur opa.eu/legal-content/EN/TXT/?uri=ecli%3AECLI%3AEU%3AC%3A2013%3A671.

EUR-Lex. (2016). *Regulation (EU) 2016/679 of the European Parliament and of the Council on the protection of natural persons with regard to the processing of personal data and on the free movement of such data, and repealing Directive 95/46/EC (General Data Protection Regulation).* https://eur-lex.europa.eu/eli/reg/2016/679/oj. Paragraphs 1–88. Accessed January 23, 2021.

EUR-Lex. (2018). *Directive (EU) 2018/2001 of the European Parliament and of the Council on the promotion of the use of energy from renewable sources.* https://eur-lex.europa.eu/legal-content/ EN/TXT/?uri=uriserv:OJ.L_.2018.328.01.0082.01.ENG. Accessed January 17, 2021.

EUR-Lex. (2019). Directive (EU) 2019/944 of the European Parliament and of the Council on common rules for the internal market for electricity and amending Directive 2012/27/EU. https:// eur-lex.europa.eu/legal-content/EN/TXT/?uri=CELEX%3A32019L0944. Accessed January 17, 2021.

EUR-Lex. Regulation (EC) No 1049/2001. https://eur-lex.europa.eu/legal-content/EN/ALL/?uri= CELEX%3A32001R1049. Accessed January 17, 2021.

European Commission. (2001). *European governance a white paper.* https://ec.europa.eu/commis sion/presscorner/detail/en/DOC_01_10. Accessed January 15, 2021.

European Commission. (2015). *A fully-integrated internal energy market*. https://ec.europa.eu/ commission/priorities/energy-union-and-climate/fully-integrated-internal-energy-market_en. Accessed January 23, 2021.

European Commission. (2016). *Clean energy for all Europeans—unlocking Europe's growth potential*. https://ec.europa.eu/commission/presscorner/detail/en/IP_16_4009. Accessed January 8, 2021.

European Commission. (2019). *Clean energy for all Europeans package*. https://ec.europa.eu/ene rgy/topics/energy-strategy/clean-energy-all-europeans_en#energy-performance-in-buildings. Accessed January 23, 2021.

Frontier Economics. (2015). *Scenarios for the Dutch electricity supply system*. https://www. frontier-economics.com/media/1046/20160415_scenarios-for-the-dutch-electricity-supply-sys tem_frontier.pdf. Accessed January 7, 2021.

Geothermie, S. P., DAGO, S. W., & EBN. (2018). *Masterplan Aardwarmte in Nederland—Een brede basis voor een duurzame warmtevoorziening*. https://www.ebn.nl/wp-content/uploads/2018/05/ 20180529-Masterplan-Aardwarmte-in-Nederland.pdf. Accessed January 22, 2021.

Gonzalez, C., Raya, S., & Elizabeth, W. (2018). *Energy justice: US and international perspectives*. Edward Elgar Publishing.

Hancher, L., Larouche, P., & Lavrijssen, S. (2004). Principles of good market governance. *Journal of Network Industries, XLIX*(2), 355–389.

Heffron, R., & McCauley, D. (2017). The concept of energy justice across the disciplines. *Energy Policy, 105*, 658–666.

Heffron, R., McCauley, D., Stephan, Hannes, Jenkins, K., & Rehner, R. (2016). Energy justice: A conceptual review. *Energy Research & Social Science, 11*, 174–182.

Hoogervorst, N. (2017). *Toekomstbeeld Klimaatneutrale warmtenetten in Nederland*. Planbureau voor de Leefomgeving. https://www.pbl.nl/sites/default/files/downloads/pbl-2017-toekom stbeeld-klimaatneutrale-warmtenetten-in-nederland-1926_1.pdf. Accessed January 21, 2021.

Huygen, A., & Akerboom, S. (2020). Geef gemeenten de vrijheid om innovatieve warmtenetten toe te staan. *ESB*, pp. 2–5. https://repository.tno.nl//islandora/object/uuid:01101755-1fb1-4152-99ab-d2feb46470e6. Accessed January 21, 2021.

Huygen, A., Beurskens, L., Menkveld, M., & Hoogwerf, L. (2019). *Wat kunnen we in Nederland leren van warmtenetten in Denemarken?* https://repository.tno.nl//islandora/object/uuid:1cf 1daff-7dee-443c-819d-14f543b4a2da. Accessed January 18, 2021.

Huygen, A., Lavrijssen, S., de Vos, C., de Wit J. (2011). *De bescherming van de consument op grond van de Warmtewet: een onderzoek naar de juridische en economische gevolgen van de gewijzigde Warmtewet voor de positie van de consument, Report*. https://dare.uva.nl/search?ide ntifier=418b8070-0830-43ea-9d34-ff20e38808e0. Accessed January 22, 2021.

IEA. (2017). *Digitalization & energy*. https://www.iea.org/reports/digitalisation-and-energy. Accessed January 23, 2021.

Jenkins, K. (2019). Energy Justice, energy democracy and sustainability: Normative approaches to the consumer ownership of renewables. In J. Lowitzsch (Ed.), *Energy transition: Financing consumer co-ownership in renewables* (pp. 79–97). Palgrave Macmillan.

Jenkins, K., McCauley, D., Heffron, R., Stephan, H., & Rehner, R. (2016). Energy justice: A conceptual review. *Energy Research & Social Science, 11*, 174–182.

Klimaatakkoord. (2019). https://www.klimaatakkoord.nl/binaries/klimaatakkoord/documenten/ publicaties/2019/06/28/klimaatakkoord/klimaatakkoord.pdf. Accessed January 7, 2021.

Koop, C., & Hanretty, C. (2018). Political independence, accountability, and the quality of regulatory decision-making. *Comparative Political Studies, 51*, 38–75, referring to Hanretty, C., & Koop, C. (2013). Shall the law set them free? The formal and actual independence of regulatory agencies. *Regulation and Governance, 7*(2), 195–214.

LaBelle, M. (2017). In pursuit of energy justice. *Energy Policy, 107*(C), 615–620.

Larouche, P. (2014). *CERRE, Code of Conduct and Best Practices for the setup, operations and procedure of regulatory authorities*. https://cerre.eu/wp-content/uploads/2014/05/140507_ CERRE_CodeOfCondAndBestPractRegulat_Final.pdf. Accessed January 16, 2021.

Larouche, P., Hanretty C., & Reindl A. (2012). *Independence, accountability and perceived quality of regulators—a CERRE study*. https://pure.uvt.nl/ws/portalfiles/portal/6846095/120306_Indepe ndenceAccountabilityPerceivedQualityofNRAs_1_.pdf. Accessed January 16, 2021.

Lavrijssen, S. (2006). *Onafhankelijke mededingingstoezichthouders, regulerende bevoegdheden en de waarborgen voor good governance*. Boom Juridische uitgevers.

Lavrijssen, S. (2018). *Een integrale Energiewet*. https://pure.uvt.nl/ws/portalfiles/portal/28119756/ 23_integrale_energiewet.pdf. Accessed January 5, 2021.

Lavrijssen, S. (2019). *Independence, regulatory competences and the accountability of national regulatory authorities in the EU*. OGEL 1.

Lavrijssen, S., & Ottow, A. (2011). The legality of independent regulatory authorities. In L. Besselink, F. Pennings, & S. Prechal (Eds.), *The eclipse of legality* (pp. 73–97). Kluwer Law International.

Lavrijssen, S., & Vitez, B. (2015). Principles of good supervision and the regulation of the Dutch drinking water sector. *Competition and Regulation in Network Industries, 16*(3), 219–259.

Lavrijssen, S., & Carrillo, A. (2017). Radical prosumer innovations in the electricity sector and the impact on prosumer regulation. *Sustainability, 9*(7), 1–21.

Lavrijssen, S., & Huygen, A. (2013). De warmteconsument in de kou: een juridische en economische analyse van de positie van de warmteconsument. *Sociaal economische wetgeving*, pp. 248–268.

Lavrijssen, S., & Ottow, A. (2012). Independent supervisory authorities: A Fragile concept. *Legal Issues of Economic Integration, 39*(4), 419–446.

Lavrijssen, S., & Vitez, B. (2020). The energy transition: Democracy, justice and good regulation of the heat market. *Energies, 13*(5), 1–24. https://doi.org/10.3390/en13051088.

Lindt, A. (2013). *Renewable energy and liberalisation of energy markets: The European experience with third party access to the grid*.

Lund, H., Østergaard, P. A., Chang, M., Werner, S., Svendsen, S., Sorknæs, P., et al. (2018). The status of 4th generation district heating: Research and results. *Energy, 164*, 147–159.

Lund, H., Werner, S., Wiltshire, R., Svendsen, S., Thorsen, J. E., Hvelplund, F., et al. (2014). 4th Generation district heating (4GDH) integrating smart thermal grids into future sustainable energy systems. *Energy, 68*, 1–11.

McCauley, D., Heffron, R., Stephan, H., & Jenkins, K. (2013). Advancing energy justice: The triumvirate of tenets and systems thinking. *International Energy Law Review, 32*(3), 107–116.

McGowan, F. (2001). Consumers and Energy liberalization. In D. Geradin (Ed.), *The liberalization of electricity and natural gas in the European Union* (pp. 63–80). Kluwer Law International.

Mendes, J. (2011). *Participation in EU rulemaking. A rights-based approach*. Oxford University Press.

Mendonça, M., Lacey, S., & Hvelplund, F. (2009). Stability, participation and transparency in renewable energy policy: Lessons from Denmark and the United States. *Policy and Society, 27*(4), 379–398. https://doi.org/10.1016/j.polsoc.2009.01.007.

Ministerie van Economische Zaken en Klimaat. (2016). *Energieagenda—Naar een CO$_2$- arme energievoorziening*. https://www.rijksoverheid.nl/documenten/rapporten/2016/12/07/ea. Accessed January 22, 2021.

Ministerie van Economische Zaken en Klimaat. (2019). *Kamerbrief over warmtewet 2.0*. https://www.rijksoverheid.nl/documenten/kamerstukken/2019/02/14/kamerbrief-over-war mtewet-2.0. Accessed January 21, 2021.

Ministerie van Economische Zaken en Klimaat. (2020). *Wet collectieve warmtevoorziening*. https:// www.internetconsultatie.nl/warmtewet2. Accessed January 21, 2021.

Morris, C., & Arne, J. (2016). *Energy Democracy—Germany's Energiewende to Renewables*. Palgrave Macmillan.

Natuur & Milieu. (2018). *Warmtenetten in de energietransitie—verkennend onderzoek naar knelpunten op basis van interviews met Zuid-Hollandse gemeenten*. https://www.natuurenmilieu. nl/wp-content/uploads/2018/08/2018-Paper-warmte-Zuid-Holland-7-augustus.pdf. Accessed January 21, 2021.

NRC Handelsblad. (2019). *Aardwarmte is de duurzame energie van de toekomst*. https://www.nrc.nl/nieuws/2019/06/14/een-waterput-om-je-huis-te-verwarmen-a3963783. Accessed January 22, 2021.

OECD. (1996). *Regulatory reform and innovation*. https://www.oecd.org/sti/inno/2102514.pdf. Accessed January 23, 2021.

OECD. (2012). *Recommendation of the council on regulatory policy and governance*. https://www.oecd.org/gov/regulatory-policy/49990817.pdf. Accessed January 18, 2021.

OECD. (2014). *The governance of regulators*. https://www.oecd-ilibrary.org/governance/the-gov ernance-of-regulators_9789264209015-en. Accessed January 8, 2021.

OECD. (2016). *The governance of regulator —being an independent regulator*. https://www.oecd.org/publications/being-an-independent-regulator-9789264255401-en.htm. Accessed January 16, 2021.

OECD. (2017). *The governance of regulators—independence of regulators and protection against undue influence*. https://www.oecd.org/regreform/independence-of-regulators.htm. Accessed January 16, 2021.

Omgevingswet. (2019). *Herleidbare geconsolideerde versie Omgevingswet*. https://www.omg evingswet.net/documenten/107-herleidbare-geconsolideerde-versie-omgevingswet. Accessed January 22, 2021.

Ottow, A. (2015). *Market & competition authorities- good agency principles*. Oxford University Press.

Overlegtafel Energievoorziening (2015). *Nieuwe Spelregels Voor Een Duurzaam En Stabiel Energiesysteem*. https://www.netbeheernederland.nl/_upload/Files/Rapport_overlegtafel_energi evoorziening_90.pdf. Accessed January 7, 2021.

Parag, Y., & Sovacool, B. K. (2016). Electricity market design for the prosumer era. *Narure Energy, 1*, 1–6.

Pellegrini, G., Pirni, A., & Maran, S. (2020). Energy justice revisited: A critical review on the philosophical and political origins of equality. *Energy Research & Social Science, 59*, 101310. https://doi.org/10.1016/j.erss.2019.101310.

Portuese, A. (2011). The principle of subsidiarity as a principle of economic efficiency. *Columbia Journal of European Law, 17*, 231–261.

Prechal, S., & De Leeuw, M. (2008). Transparency: A general principle of EU law? In U. Bernitz, Joakim Nergeliu, & Cecilia Cardner (Eds.), *General principles of EC law in a process of development* (pp. 201–238). Kluwer Law International.

Prosser, T. (2006). Regulation and social solidarity. *Journal of Law and Society, 33*(3), 364–387.

Rijksoverheid. (2021). *Kan ik zelf een energieleverancier kiezen bij blokverwarming of stads-verwarming?* https://www.rijksoverheid.nl/onderwerpen/energie-thuis/vraag-en-antwoord/kan-ik-een-energieleverancier-kiezen-bij-blokverwarming-of-stadsverwarming. Accessed January 18, 2021.

Ropenus, S., & Henrik, J. (2015). *A snapshot of the Danish energy transition: Objectives, markets, grid, support schemes and acceptance*. Agora Energiewende.

Schepers, B. (2009). *Warmtenetten in Nederland, report*. https://www.ce.nl/publicaties/downlo ad/782. Accessed January 7, 2021.

SiRM. (2019). *Tariefregulering warmtebedrijven voor kleinverbruikers*. https://www.sirm.nl/docs/Publicaties/20190131-SiRM-def-rapport-Regulering-kleinverbruikers-warmtenetten.pdf. Accessed January 23, 2021.

Sovacool, B., & Dworkin, M. (2015). Energy justice: Conceptual insights and practical applications. *Applied Energy, 142*, 435–444.

Sovacool, B., Burke, M., Baker, L., Kotikalapudi, C., & Wlokas, H. (2017). New frontiers and conceptual frameworks for energy justice. *Energy Policy, 105*, 677–691.

Sovacool, B., & Dworkin, M. (2015). Energy justice: Conceptual insights and practical applications. *Applied Energy, 142*, 435–444.

Steyn, E. (2014). Electricity market restructuring: Perspective from abroad. *SA Mercantile Law Journal, 26*(3), 606–650.

The World Bank. (1992). *Governance and development.* http://documents1.worldbank.org/curated/en/604951468739447676/pdf/multi-page.pdf. Accessed January 15, 2021.

Tieben, B., & Van Benthem, M. (2018). *Belang bij splitsing in de warmtemarkt - Effecten van splitsing op publieke belangen in de warmtemarkt. Rapport nr. 2018-98.* https://spa-sg.nl/wp-content/uploads/2019/06/SEO-Belang-bij-splitsing-in-de-warmtemarkt-2018.pdf. Accessed January 21, 2021.

United Nations. (2015). *Adoption of the Paris Agreement.* https://www.unfccc.int/resource/docs/2015/cop21/eng/l09.pdf. Accessed January 5, 2021.

Van Veelen, B., & Van Der Horst, D. (2018). What is energy democracy? Connecting social science energy research and political theory. *Energy Research & Social Science, 1*(46), 19–28.

Zilman, D., Lee, G., LeRoy, P., & Martha, R. (2018). *Innovation in energy law and technology: Dynamic solutions for energy transitions.* Oxford Scholarship Online.

Connecting Technology and Society

Enabling Public Participation in Shaping the Inclusive Energy Transition Through Serious Gaming—Case Studies in India

Bharath M. Palavalli, Sruthi Krishnan, and Yashwin Iddya

Abstract To create holistic plans for equitable access to energy and to create sustainable transition pathways, stakeholder consultation and engagement processes are essential. In India, the planning process for energy has challenges that range from legacy processes, increasing energy demand to fuel growth, pressures arising from competing (as well as new and old) technologies, to varying goals for all the stakeholders. We categorize these factors as institutional structures, geopolitical, environmental, technical, social, and monetary factors. To ensure a vision for a collective future and a coherent plan for energy, it is important that the processes enable participation and allow for co-ordination and interaction to strengthen dialogue. Processes should capture intangibles and include slack for events such as pandemics, which are no longer treated either as externalities or once-in-a-lifetime events. In this chapter, we give two examples of serious games as tools to address these challenges in the context of planning. The first example is of a game created for bureaucrats, decision-makers in the government, and private energy companies to plan collectively and compare results from various plans for energy expenditure in India. In the second case, the game aids transportation planning in urban India, which requires additional effort to ensure a transition to equitable access to energy. Using results from the game sessions, we illustrate how such methods can bridge gaps in energy planning in the diverse and challenging context of India.

Introduction

Energy is and will be the main driver of the world economy. The energy sector however is also a big contributor to global warming. In 2008, the energy sector accounted for more than 80% of anthropogenic greenhouse gas emissions for

B. M. Palavalli (✉) · S. Krishnan · Y. Iddya
Fields of View, https://www.fieldsofview.in, India
e-mail: bharath@fieldsofview.in

© The Author(s) 2021
M. P. C. Weijnen et al. (eds.), *Shaping an Inclusive Energy Transition*,
https://doi.org/10.1007/978-3-030-74586-8_10

Annex 1[1] countries (Akpan & Akpan, 2012). In the case of India in 2015, nearly 58% of the greenhouse gas emissions were attributed to the energy sector (TNN, 2015). In order to effectively tackle climate change a transition of the energy sector towards renewable sources and adoption of increased energy efficiency is required. Additionally, access to modern energy services is fundamental to fulfilling basic social needs, driving economic growth and fuelling human development (Gaye, 2007). The role of energy planning is hence crucial in providing pathways for the transition.

However, there is a lack of consensus on the exact definition of energy planning (Cajot et al., 2017). Thery and Zarate (2009) define energy planning as "determining the optimal mix of energy sources to satisfy a given energy demand". According to Keirstead et al. (2012), the purpose of energy planning is to balance the spatially localized energy supply and demand of a given area. However this encompasses a variety of processes, energy carriers and technologies that are rarely managed together, as should be, for example, supply, conversion, storage and transportation technologies (Loken, 2007). Also, energy planning is generally not a systematically established institution within administrative departments (Cajot et al., 2017). Energy planning often is spread across multiple stakeholders, each of whom might view the planning and issues differently. The role of energy planning, apart from the technical sphere, needs to include perspectives from social-economic and environmental spheres to align with the sustainable development agenda. The complex interactions in these three spheres, different stakeholders involved in planning, and difficulty in defining it makes it a 'wicked problem' (Cajot et al., 2015,2017; Thollander et al., 2019).

The expression 'wicked problem' was coined in 1973 by Rittel and Webber (1973). According to them *"The kinds of problems that planners deal with—societal problems—are inherently different from the problems that scientists and perhaps some classes of engineers deal with. Planning problems are inherently wicked."* Wicked problems are complex problems, which usually involve multiple stakeholders with their own worldviews, leading to variable definitions of the problem, which can sometimes be contradictory (Cajot et al., 2015; Garcia et al., 2016; Thollander et al., 2019). Such problems show resistance to resolution because of incomplete, contradictory, and changing requirements (Coulton et al., 2014; Thollander et al., 2019). Due to the uncertainty and ambiguity that wicked problems involve, technical analysis is unlikely to provide a final resolution (Thollander et al., 2019). Therefore, we require methods to effectively address the nature of wicked problems. According to Conklin (2018), two things must happen to make progress on wicked problems. The first is for stakeholders to collaboratively gain a shared understanding of a problem, as opposing stakeholders usually do not even agree on what the problem is. The second is for opposing stakeholders to have dialogue and a shared commitment to alleviating the problem.

Games are well-suited to communicating a shared understanding of a problem between different stakeholders because they allow users to experiment with potential

---

[1] The countries that are included in Annex I of the United Nations Framework Convention on Climate Change as amended on 11 December 1997 by the 12th Plenary meeting of the Third Conference of the Parties in Decision 4/CP.3.

solutions in a safe setting and generate their own mental frames for how it works. Additionally, since the cost of failure in games is low, players may be emotionally more capable of trying out different ideologies (Swain, 2007). According to Garcia et al. (2016), games are a powerful tool for engaging users, letting them explore the complexities of a system, and giving them the opportunity to deal with wicked, ill-defined problems in a safe and fun environment. This has led to games being used in different parts of the world to plan in multiple domains such as city-plans, transport, energy, electricity, resource management and participatory planning. In this chapter, we first explore the complexity and wickedness of the energy policy scenario in India. We then demonstrate, through means of case studies, how serious games can aid in resolving those issues in the Indian context.

Energy Planning Scenario in India

Considering the geographic size of India, its population and its increasingly growing energy needs, energy policy in India is complex in nature as can be explained due to the following factors.

Institutional Structure

There is lack of an integrated energy policy structure across India (Planning Commission, 2006). The Indian Energy Policy sector has multiple key players, both at the centre and the state government levels, with functions distributed as per 7th schedule of the Indian Constitution. At the central level, there are six ministries under the Government of India which are responsible for energy policy design and implementation. Each institution involved has its own mandate and objectives. For instance, DAE's objective is to increase the nuclear power generation capacity and MoP's objective to add coal generation capacity. Each ministry is interested in promoting the generation technologies they are responsible for (Ahn and Graczyk, 2012). Furthermore, there are private sector industries involved in electricity generation and research and development (refer Fig. 1 in TERI Energy Data Directory Institute, 2010). Also, there are other sectors like road transport, shipping, housing, etc., which impact the energy usage in the country each of which are under a different ministry, and are not considered part of energy planning in India. This structure leads to siloed functioning, and can make achieving objectives of energy planning difficult in India.

Geopolitical Factors

Though one of the policy objectives in the energy policy framework is energy security and independence, India is still highly dependent on fuel imports for electricity generation. India imports nearly 75% of its energy need from oil (Babajide, 2018).

Geopolitical factors play a key role in some of the energy choices. For example, increasing natural gas plants capacity would mean increased dependence on imports, which would leave the country vulnerable to international price shocks (Planning Commission, 2006).

Environmental Factors

Though India's greenhouse gas emissions (GHG) are less than the global average, the sustainable development policy direction makes it important to contain emissions (Planning Commission, 2006). Given the relationship between economic growth and sustainable development, there is both internal and international pressure to keep the GHG emissions under a certain level. This increases the pressure to use cleaner fuels and invest in research and development.

Technical Factors

Some legacy technologies have associated technical issues that limit their capacity; such as high variation in grids resulting from intermittent generation from wind and solar power plants (Sharma et al., 2018). To address the issue, additional capacity that can act as balancing load needs to be installed.

Social Factors

One of the primary objectives in energy policy is Universal Energy Access (Ahn & Graczyk, 2016). However, this has been difficult to realise due to capacity shortage and affordability issues. This makes producing electricity at very low costs one of the key constraints. Similarly ensuring availability of modern fuels for cooking and heating at affordable prices in rural areas has been a key challenge. Additionally, some of the generation technologies have high societal costs. For instance, large hydro-electric projects displace millions of people, and affect the surrounding ecosystem as well.

Monetary Factors

India's economy is growing and this needs a matching growth in energy capacity. Government have a limited budget to work with, making it hard to achieve the desired energy targets (Institute & "TERI Energy Data Directory & Yearbook", 2010).

Serious Games—Role in Shaping an Inclusive Energy Transition

As discussed before, games can become an important methodology to address the wickedness of energy planning. They enable dialogue between diverse stakeholders, help in understanding the complexity and capture the intangible needs and preferences of people, which traditional planning methodology neglects. We demonstrate this through case studies of two games we have built. First is the Indian Energy Game, where the players are involved in designing the electricity mix for India across 2 five-year planning cycles. Second is Transport Trilemma, where players play the role of a Bus Transport Corporation for the city of Bangalore, taking planning decisions relating to various aspects of bus transport.

Case Study: Indian Energy Game

The Indian Energy Game was designed as a learning tool to help participants learn about the complexity involved in designing energy policy in India, by allowing them to experience the policy making process (Hoysala et al., 2013). The game was targeted at the public at large, and for people with a working knowledge of the energy sector.

About the game:

The game has three roles modelled after the institutional structure for electricity policy in India:

1. Ministry of Power (MoP)[2]: In the game, MoP controls the decisions about Coal based thermal plants, Hydroelectric Power plants and Natural Gas based thermal plants.
2. Department of Atomic Energy (DAE)[3]: In the game, DAE controls the decisions about Nuclear Energy.
3. Ministry of New and Renewable Energy (MNRE)[4]: In the game, MNRE controls decisions about Solar and Wind Energy.

In the first round, the participants play the 12th Five-Year plan (2012–2017) and in the second round, they play the 13th Five-Year plan (2017–2022).[5] Throughout the course of the game, the participants are provided with messages, which describe

---

[2]Ministry of Power, https://powermin.nic.in/en/content/about-ministry, last accessed on 28th April 2019.

[3]Department of Atomic Energy, https://dae.nic.in/?q=node/634, last accessed on 28th April 2019.

[4]Ministry of New and Renewable Energy, https://mnre.gov.in/history-background, last accessed on 28th April 2019.

[5]Until the year 2014, from the year 1951, India followed the process of five-year integrated plans at the central, national level, https://niti.gov.in/planningcommission.gov.in/docs/plans/planrel/fiveyr/welcome.html, last accessed on 3rd January 2020.

various constraints that the participants experience. For example, the message "90% of Natural Gas is imported" is provided to the MoP to inform them about that the availability of natural gas is dependent on political calm in the region. The message "Hydro projects will displace people, for which the rehabilitation costs are high" is provided to the MoP to describe the social cost of building large hydroelectric projects. Technical factors such as unstable networks due to wind energy generation are also introduced through messages.

The participants have three objectives they need to satisfy as a group. They need to

- add a capacity of 76,000 MW in the first round,
- maintain the price of generation per kWh at Rs. 3,
- and maintain CO_2 emission levels at 395 Million Tonnes of CO_2.

Each of the ministries has a pre-defined budget to meet these targets. The objectives for the second round are dependent on the players' performance in the first round.

Enabling Participation

The energy game was designed for a wide audience ranging from people with little or no knowledge of the energy sector, to experts working in the sector. This design allows making energy policy more accessible to people. It helps a citizen to better understand the complexity of energy policy making, hence enabling informed participation. We have played the India energy game with widely varying sets of participants. In each game session, we documented the background of the participants, captured interactions between the players throughout the course of the game, their responses to the messages provided to them and the debrief sessions. We have 9 game sessions and 6 play-tests of the Indian Energy Game. These nine sessions had the following mix of participants:

- Session 1 and 7: The participants had little or no knowledge about energy policy design in India.
- Sessions 2 and 3: The team was a mixture of people who had a working knowledge of energy policy design and its complexity in India, people who worked in the energy sector and people who had little or no knowledge about energy policy design. Session 3 had a member from the Planning Commission as a participant.
- Sessions 4, 5 and 6: The participants had a working knowledge of energy policy design and its complexity in India. Furthermore, they had a bias towards clean energy and were staunchly against the use of nuclear power.
- Sessions 8 and 9: The participants had a working knowledge of energy policy design in India and were from the Indian Administrative Services (IAS).

Irrespective of the background of the participants the 9 game sessions had the following common feedback from participants shown in Table 1.

Table 1 Observations common across 9 game sessions (Hoysala et al., 2018)

Participant's learning and experiences	Observation from game play data
High life-cycle cost of solar energy	The teams reduce the amount of solar capacity added over the course of the game
High reliance on coal based energy	The teams begin the game with a mixture that results in a high cost of generation. They gradually reduce the mean cost of generation by relying on coal based energy sources
Non-availability of inexpensive hydroelectricity	The teams begin with an average of 25% hydroelectricity. As constraints about hydroelectricity are introduced in the course of the game, the average share of inexpensive hydroelectricity decreases

Hence games can be used as a medium to make complex policy challenges understandable by people, which has potential to enhance their participation in policy making and have an informed discussion.

Coordination Among Players

In order to successfully win the game, coordination among players is important. As each player is in charge of different technologies assigned to their respective Ministries/Department, there are trade-offs involved. For example, while solar power has near zero GHG emissions, the cost of generation is high; whereas coal power has lower cost of generation but higher GHG emissions. As both coal and solar are under the control of different players, players need to break out from their silos and coordinate with other players for arriving at the electricity mix that helps achieve the objectives in the game. During the game sessions it was found that over time the coordination among different players improved, and the teams reached closer to the objectives in the game. In Fig. 1 and Fig. 2 you can see the average cost of generation and average emissions, respectively, across different time bins. The values are averaged across 9 game sessions. As time progresses you can see the convergence with the objective set in the game both for the cost of generation and emissions.

Enabling Dialogue

In energy planning, as discussed earlier, there are multiple stakeholders, each of them with their own perspectives and objectives. Often, these objectives are contradictory and lead to disjoint planning. To shape an inclusive energy transition, there is a need to increase dialogue among different stakeholders, in order to arrive at a shared understanding of the future and therefore work together towards it. For this

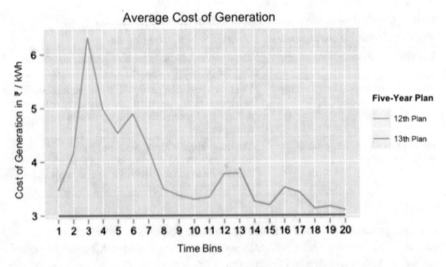

Fig. 1 Average cost of generation (across 9 sessions) (Hoysala et al., 2018)

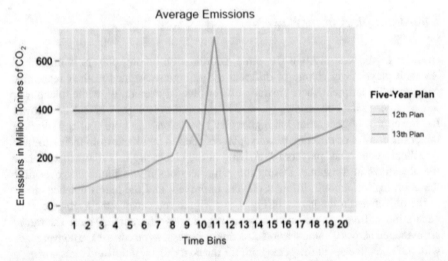

Fig. 2 Average GHG emissions (across 9 sessions) (Hoysala et al., 2018)

to happen, exposure to different perspectives is a starting point. Additionally, due to the different backgrounds and ideologies of the stakeholders, there is an added impact on policy-planning, for example, an anti-nuclear stance will lead to an energy mix devoid of nuclear energy, irrespective of its benefits. Games enable the players to explore the consequences of their choices in a low-cost manner, hence, enabling better engagement of the pros and cons of each policy stance. In the game sessions 4, 5 and 6, the players had a bias towards clean energy and were staunchly against the use of nuclear energy. Their feedback after the game session and their decisions during the game is discussed in Table 2.

Despite the bias towards clean energy, the game enabled participants to understand the social side of the issue, mainly the affordability of energy. In India, the high reliance on coal based power is due to it being among the cheapest sources of power and also for geo-political reasons, as India has indigenous reserves of coal, whereas it

Table 2 Experience and feedback of participants in game session 4, 5 and 6 (Hoysala et al., 2018)

Game play details	Participants' learning and experiences	Observations from the game play data
Session 4	Tried a strategy of developing an energy mixture without coal Game helped them understand better the consequences of the same	Refer Fig. 3. The team did not use coal based power, and eventually failed to meet the energy requirements of the country in the game
Session 5	The high financial and lifecycle emission consequences of installing solar power became evident during the course of the game	Refer Fig. 4. The team began with a high percentage of solar power, which reduced the available budget due to its cost
Session 6	The game helped understand that the environmental, social, monetary, institutional and technical factors determining energy policy cannot be isolated The game helped them understand the environmental and societal costs of an energy mixture without nuclear energy	Refer Fig. 5. The graphs show that the teams used no nuclear power but generated 75% of the power through coal, thereby potentially increasing emissions and decreasing coal reserves

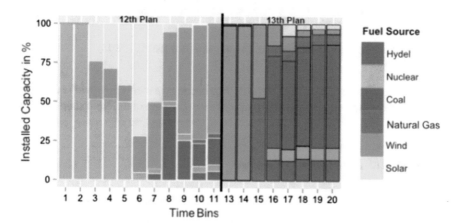

Fig. 3 Electricity mix in session 4 throughout the game (Hoysala et al., 2018)

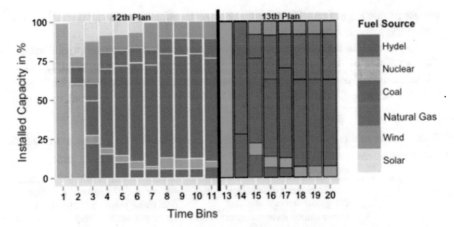

Fig. 4 Electricity mix for session 5 throughout the game (Hoysala et al., 2018)

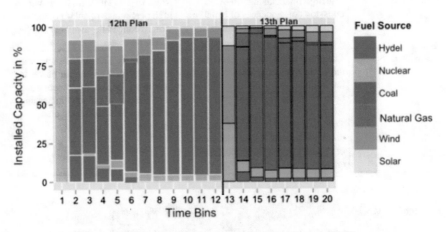

Fig. 5 Electricity mix for session 6 throughout the game (Hoysala et al., 2018)

is reliant on petroleum imports. The game thus becomes a medium to bridge different viewpoints for the same issue, which enables dialogue among different stakeholders.

Case Study: Transport Trilemma

Now let us move away from the electricity sector, to the transportation sector, where we have developed a game called Transport Trilemma. Transport Trilemma is a novel scenario-generation transport planning game designed for the context of public bus transportation in the city of Bangalore. Bangalore Metropolitan Transport Corporation (BMTC), the public bus transport corporation in Bangalore, is considered to be the "lifeline of Bangalore" (Bangalore Mirror Bureau, 2017). Its buses deliver close to 5 million passenger trips everyday with more than 2000 routes and over 6000 buses (BMTC). Yet, various commuter surveys and feedback suggest that fares are expensive, wait times are long, schedules are unreliable besides numerous other commuting challenges with the existing service. For many people, especially the urban poor, the bus service forms the only means of transport apart from walking, due to unaffordability of other means (Bangalore Mirror Bureau, 2017). Transport Trilemma is a tool built for the planners/policymakers of BMTC with an objective to explore the creation of transport plans that are more inclusive and responsive to the needs of all stakeholders as a part of our Joint Road Forward project.[6]

About the game:

The goal of the game is to provide a participatory, alternative approach to planning and assessment of public bus transportation plans. This gaming simulation is based on a model that incorporates data that conventional methods rely on and also allows planners to 'play with' intangible data such as needs, preferences and priorities that conventional methods fail to capture. The objective is to use the collected results and strategies in the game in an agent-based simulation to understand and consider the different trade-offs while making decisions. This approach also allows planners to experiment with different choices, explore alternative planning scenarios, and consider different trade-offs through this combination of methods which will also provide a low-cost and risk-free environment to test new strategies.

In the game, players assume the role of BMTC and are responsible for ensuring that the operation of buses is profitable while also increasing its ridership in the city by bringing in new commuters from different income segments in Bangalore. Annual plans are created for the period 2018–2020 to achieve the overarching targets set for annual ridership and annual gross revenue using BMTC's operations and maintenance parameters. The historic operational details and parameter values for the years 2005–2017 are provided as information guides for the players to plan and set targets. Throughout the course of the game, constraints are introduced in the form of "messages" as a part of the game experience; constraints that planners encounter in the real world.

Capturing the Intangible

We conducted the game with two different sets of players. In each game session, we documented the background of the participants, captured interactions between the players throughout the course of the game, and their responses and feedback from the debrief sessions.

---

[6]Details available at last accessed on 28th April 2019 https://fieldsofview.in/projects/joint-road-for ward/, last accessed on last accessed on 28th April 2019.

- Session 1: Members from BMTC, the public bus transport corporation in Bangalore
- Session 2: Members from BBPV (Bangalore Bus Prayaanikara Vedike), a commuters' group which conducts public discussions on urban mobility issues covering accessibility and affordability of public transport services.

While the context of the game was the same across both sessions, yet the outcomes of the game are different, due to different needs and priorities of the players. Some of the results of the sessions are briefly discussed in Table 3.

The intangible needs and preference of the stakeholders become visible from their choices in game. From the results of the game sessions, BMTC places a greater priority on ridership and revenue, whereas BBPV places higher priority on accessibility and affordability. These game interactions generate important information that cannot be ignored in planning. Plans need to take cognizance of different stakeholder needs in order to be inclusive and responsive to them. Even when the same context is provided to different stakeholders, we can clearly see how different outcomes are generated. Both the stakeholders considered different factors and goals to guide their decision-making, which results in different planning outcomes. The needs, preferences and priorities differ between stakeholders due to which the plans and their outcomes also change. Transport trilemma helps to sensitise stakeholders to the needs and priorities of other stakeholders involved and makes them aware of the impact of decision-making on each other. Considering that the resources are limited, various trade-offs get introduced into the decision-making, which leads to prioritisation. It becomes necessary for planners to be wary of all the various trade-offs and their impacts before they take any decision and prioritise so that plans can be made more inclusive, participatory and responsive to the needs of different people affected by it. Tools like Transport Trilemma allow for participation of different stakeholders, enable dialogue, explore alternatives and sensitise different stakeholders to each other's needs, priorities and constraints while also making them holistically aware of the system they are operating in.

Table 3 Results from game sessions of Transport Trilemma

Session 1	Session 2
Strategies were based on achieving high ridership and high revenues for BMTC	Strategies were mostly based on increasing accessibility and affordability of BMTC's different services to commuters
Decisions were based on the assumption that an increase in the number of buses on road will result in increased ridership	Preference for routes and schedules over increasing the fleet strength, though the number of buses was also increased
Targets were incrementally increased across the years without any significant change	Ambitious targets were set for all planning years signalling a deviation from historical trend
Was more interested in those routes which were high in revenue for BMTC while others were neglected	Fares were either kept nominally low or even reduced year on year to make BMTC more affordable

Conclusion

While the transition towards clean and renewable sources of energy is inevitable if we want to ensure sustainable development for all, the challenges posed by the wickedness of the energy planning problem makes the existing methodology we use to solve the problem and improve public participation inadequate. Games offer solutions to address the wickedness of the problem. Through the case studies of Indian Energy Game and Transport Trilemma, it was shown how serious games are being used to address the challenges in energy planning in India. Specifically, we have described with case studies in India how games help in:

- Enabling Participation of citizens in policy making, by making policy accessible.
- Improving Coordination among diverse stakeholders, enabling working towards a common goal.
- Enabling Dialogue among different stakeholders, helping to understand the problem in a holistic manner.
- Capturing the intangible needs and preference of stakeholders, to make policy inclusive.

Serious Gaming acts as a complement to the existing approaches. The combination of games with conventional approaches is needed to ensure an inclusive energy transition.

References

Ahn, S.-J., & Graczyk, D. (2012). *Understanding energy challenges in India: Policies, players and issues*, IEA.

Akpan, U. F., & Akpan, G. E. (2012). The contribution of energy consumption to climate change: A feasible policy direction (Vol. 1), p. 13.

Babajide, N. (2018). Indian energy security status: What are the economic and environmental implications? p. 5.

Bangalore Mirror Bureau (2017, July 2). Bengaluru's Road-runner, *Bangalore mirror*.

BMTC, Data on buses-routes, fares, bus stop.

Cajot, S., Peter, M., Bahu, J.-M., Guignet, F., Koch, A., & Maréchal, F. (April 2017). Obstacles in energy planning at the urban scale. *Sustainable Cities and Society, 30*, 223–236

Cajot, S., Peter, M., Bahu, J.-M., Koch, A., & Maréchal, F. (November 2015). Energy planning in the urban context: challenges and perspectives. *Energy Procedia, 78*, 3366–3371

Conklin, J. (2018). *Wicked problem and social complexity*, Cognexus Institute.

Coulton, P., Jacobs, R., Burnett, D., Gradinar, A., Watkins, M., & Howarth, C. (2014). Designing data driven persuasive games to address wicked problems such as climate change. In *Proceedings of the 18th international academic mindtrek conference on media business, management, content & services—AcademicMindTrek '14*. Tampere, Finland, pp. 185–191.

Garcia, C., Dray, A., & Waeber, P. (January 2016). Learning begins when the game is over: using games to embrace complexity in natural resources management. *GAIA—Ecological Perspectives for Science and Society, 25*(4), 289–291

Gaye, A. (2007). Access to Energy and human development, p. 22.

Hoysala, O., Palavalli, B. M., Murthy, A., Meijer, S. (2013). Designing energy policy through the Indian energy game. In *Online proceedings of the 44th conference of international simulation and gaming association*, p. 11.

Keirstead, J., Jennings, M., & Sivakumar, A. (August 2012). A review of urban energy system models: Approaches, challenges and opportunities. *Renewable and Sustainable Energy Reviews, 16*(6), 3847–3866

Loken, E. (September 2007). Use of multicriteria decision analysis methods for energy planning problems. *Renewable and Sustainable Energy Reviews, 11*(7), 1584–1595

Planning Commission. (2006). Government of India, Integrated Energy Policy: Report of Expert Committee.

Rittel, H. W. J., & Webber, M. M. (June 1973). Dilemmas in a general theory of planning. *Policy Sciences, 4*(2), 155–169

Sharma, H., Kumar, P., Pal, N., & Sadhu, P. K. (2018). Problems in the accomplishment of solar and wind energy in India. In *Problemy Ekorozwoju*, p. 9.

Swain, C. (2007). *Designing games to effect social change*, p. 5.

The Energy Resource Institute. (2010). *TERI energy data directory and yearbook*.

Thery, R., & Zarate, P. (September 2009). Energy planning: A multi-level and multicriteria decision making structure proposal. *Central European Journal of Operations Research, 17*(3), 265–274

Thollander, P., Palm, J., & Hedbrant, J. (March 2019). Energy efficiency as a wicked problem. *Sustainability, 11*(6), 1569

TNN. (2015, December 22). Energy sector accounts for 58% of greenhouse gas emissions: Govt. In *The Economic Times*.

A Comprehensive Engineering Approach to Shaping the Future Energy System

Zofia Lukszo and Samira Farahani

Abstract The urgency to significantly reduce the impacts of climate change is felt around the globe. By signing the Paris agreement in 2016, 195 governments have agreed on a long-term goal of keeping the increase in global average temperature below 2 °C above preindustrial levels and on aiming to limit the increase to 1.5 °C. To reach these goals, major technological, organizational, and social changes in different sectors and their services are needed. To understand and steer the transition from the current energy system towards a carbon-free energy system, we propose a comprehensive engineering framework that integrates different aspects, such as technical, economic, cyber-physical, social, institutional and political, that are needed in the design of such a complex system. We explain the importance of combining different disciplines to provide comprehensive models and tools in order to support and achieve a sustainable, affordable, reliable and inclusive energy transition.

Introduction

In the coming years of energy transition the world will undergo one of the most complex technological transformations in history, which will not leave society and everyday life untouched. The challenges regarding replacing the existing fossil-based energy system by a sustainable one are not only of a technological and economic nature, but also institutional, entrepreneurial and ethical. The challenge is not only to meet our current energy needs, but also the needs of a projected 10 billion people in 2050 and to do so with low cost as well as with low-carbon and renewable energy sources.

If we have to replace the existing energy system, it is important to clarify what is meant by this term. The word 'system' is derived from Greek and means 'an aggregation of parts'. Ever since the time of ancient Greece, the term has been used

Z. Lukszo (✉) · S. Farahani
Faculty of Technology, Policy and Management, Delft University of Technology, Delft,
The Netherlands
e-mail: z.lukszo@tudelft.nl

© The Author(s) 2021
M. P. C. Weijnen et al. (eds.), *Shaping an Inclusive Energy Transition*,
https://doi.org/10.1007/978-3-030-74586-8_11

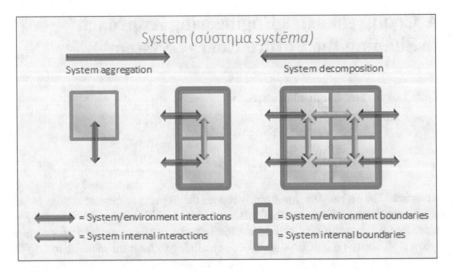

Fig. 1 Schematic representation of a system

to describe any organisation as a structured set of parts. Physical, biological, but also social and commercial systems are all examples of this.

The fact that systems can be distinguished in various disciplines and worlds makes it possible to describe the word 'system' in abstract or even in mathematical terms. It is then a combination of systems or elements that interact with each other and that can, and mostly do, also interact as a whole with their environment, see Fig. 1. In this sense a system of systems is also a system.

We can also see such composite structures in the everyday energy systems that have been designed to provide energy services to end users. If we translate such an energy system into a scheme, we may produce a sketch like this, which displays the physical system and the many subsystems and interactions it comprises, see Fig. 2. In this figure we did not include heating and cooling systems, so only a part of the physical energy system is presented here.

This large physical system is complicated enough on its own, but it is important to realise that the economic systems (involving energy markets and transactions), and the social systems (involving multiple stakeholders) are not shown in this diagram. Adding those makes this whole system much more complex.

The physical system diagram presented in Fig. 2 shows an energy system that is a collection of physical flows and conversion processes. It comprises solely passive elements that can only follow the choices of the designers, system managers and other actors. There are no smart elements, smart subsystems or active elements (whereby an active element is one that can make decisions independently). These elements are first encountered in the *cyber layer*: the part of our energy system where data is collected, the communication between the system components is accommodated, the components themselves are controlled, and the security protocols are defined and activated when necessary, see Fig. 3. This is the domain of ICT, measurement and

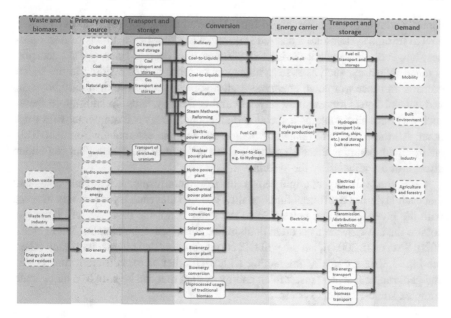

Fig. 2 Schematic representation of an energy system

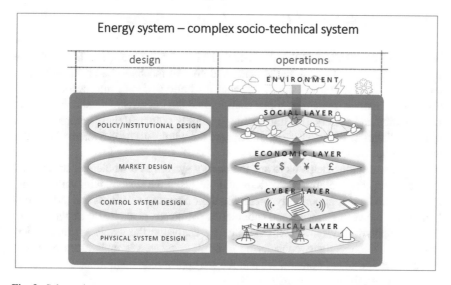

Fig. 3 Schematic representation of an energy system as a complex socio-technical system

control technology and smart algorithms. These advanced control systems calculate how to bring a system into a desired state, even in situations with lots of uncertainties; however, they cannot decide exactly what this desired state is. That decision is made by actors, such as system managers or other parties residing in the social domain. In other words, an energy system involves much more than only physical and control layers representing physical and control systems. In addition to these, such a system also contains a social and an economic layer that interact with each other. All these four interrelated layers interact with an external environment. This environment contains those elements that are not part of the system, but can either affect it or are affected by the system.

Using the term "layer" might be to some extent confusing, as it not only suggests that each layer has a different functionality, but also that each layer can only interact with the layer above and below. Since the latter is not the case, we will hereafter not use the term *layers,* as is often done in literature (see, for example, (Mittal et al., 2020; Sosa et al. 2003)), but we will talk about *domains* that interact with each other, see Fig. 3. These domains interacting with each other, across different time scales and geographical scales, result in an energy system that can be seen as a large-scale complex socio-technical system.

Systems Thinking and Complex Adaptive Systems

(Re-)designing such a complex socio-technical system calls for a coherent design involving not only physical components and control mechanisms on the one hand and energy markets on the other, but also institutions taking into account social acceptance, market dynamics, social routines and the interests of all stakeholders in the economy and society. It is impossible to quickly sketch all these complexities in a simple diagram. To perceive and understand the whole, we need Systems Thinking and a systems perspective.

Systems Thinking helps us to see the energy system as a whole, constructed by different domains operating on various scale levels in the system, with various time constants, and, most importantly including their interconnectedness and the interrelationships between elements and subsystems (Moncada et al., 2017; Senge, 1990). Only by applying Systems Thinking, i.e., by analysing the whole system, understanding interconnections, considering multiple perspectives and values of actors operating in the social domain, we can realize appropriate institutions and rules of play to support the energy transition.

Figure 4 shows schematically that in (re-)designing future energy systems a whole system perspective, including the technical, cyber, economic and social domain, the latter including the individual behaviours of actors, and their interactions with institutions as well as with the physical and cyber domains, is necessary to understand, analyse and steer this complex socio-technical energy system in the desired direction of sustainability, affordability, reliability and safety. However, knowing which domains should be taken into account is not sufficient to design possible pathways

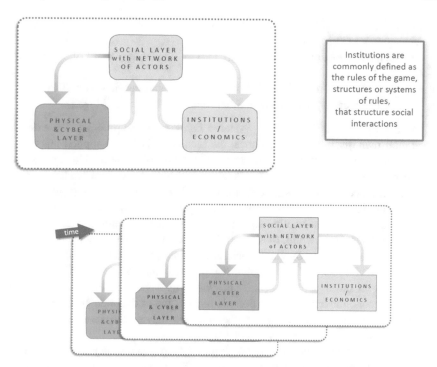

Fig. 4 Systems thinking for the (re-)design of energy systems and time dependent adaptation. The top figure shows a static socio-technical system, which should be replaced in the analysis and design by the bottom figure to be adaptable to changing conditions in the system environment

for the energy transition. Besides Systems Thinking, theories of Complex Adaptive Systems (CAS) are helpful. CAS are seen as non-linear, dynamic systems with the ability to adapt to changing conditions in the environment, and they often show unexpected (emergent) behaviour and a strong path-dependency (Amer Power Plant, 2020). The time dependent adaptation of CAS is shown schematically in the lower part of Fig. 4 where the domains and interactions constantly adapt themselves to dynamic changes in the environment, and therefore, proposed solutions should also be adaptable.

Complex Socio-Technical Systems Engineering

Many international educational programmes and research groups have made Systems Thinking a cornerstone of their research and education, and are often viewed— and consider themselves—as multidisciplinary, interdisciplinary or transdisciplinary groups. To name a few of these programs contributing to complex systems engineering education, one could mention:

- Delft University of Technology (TUDelft); Faculty of Technology, Policy and Management (TPM)
- Massachusetts Institute of Technology (MIT); Technology and Policy Program (TPP)
- Carnegie Mellon University (CMU); Engineering and Public Policy (EPP)
- Stanford University; Management Science and Engineering (MS&E).

Over the years, and initiated by Systems Thinking, these programs have developed their own discipline. For instance, the faculty of TPM has applied this to the energy and telecommunications industries and to transport systems. These programs have been started as multidisciplinary faculties with a wide variety of academic backgrounds: engineering scientists, economists, mathematicians, psychologists, physicists and ecologists, to name just a few. Just as physics has many interfaces with chemistry, biology, mathematics, etc., so have these programs countless interfaces with other disciplines. Meanwhile, these faculties are by themselves independent scientific disciplines that focus on major societal challenges with their specific social and technological complexities. We call this discipline *Complex Socio-Technical Systems Engineering*, or shortly *Comprehensive Engineering*. It is a discipline that recognises the social, economic and technical complexities of major dynamic systems, studies how these systems interact with each other and how they need to be designed and/or operated to satisfy desired states and/or values.

The question that is relevant for this chapter is: How can we deploy the scientific discipline of *Complex Socio-Technical Systems Engineering* to facilitate the inclusive energy transition? An inclusive energy transition is a shared goal of the entire international community and entails switching from fossil to renewable energy sources. This transition is needed to secure our energy supply for the future and to mitigate climate change. A consequence of the transition is that, by 2050, the share of renewable energy in the power sector will increase to 85%, compared to 25% in 2017 (IRENA, 2018). To reach this goal, countries all around the world have started changing their energy production policies to move towards more renewable energy resources.

While many countries are mainly focused on replacing coal in their energy system, some go even further and plan to replace natural gas as well. Netherlands is a good example of the latter-mentioned countries. In late January of 2018, the House of Representatives in the Netherlands passed a bill to accelerate the energy transition (Electricity Act Amendment, 2018). The new law aims to remove existing barriers to the energy transition in the Netherlands and adapt the Electricity Act and Gas Act to adequately meet the new policy targets. The most important changes involve:

- a broader experimentation provision with a reference framework for grid operators who have to choose between strengthening the grids or deploying the flexibility of market parties (so-called demand response)
- scrapping the gas connection obligation for new-built homes and for small businesses (gasless new-builds are now the norm, unless there are serious public interests that make it strictly necessary to provide a gas network).

Further, it has been decided in 2019, among other things, to convert the two oldest coal-fired power plants in the Netherlands, i.e., the Amer and Hemweg plants, by using a more sustainable fuel (Amer Power Plant, 2020; Decision of the Dutch Government, 2020). The new coal-fired plants in the Maasvlakte and Eemshaven are to undergo the same treatment, only a few years later. But that is still not enough to achieve the carbon-reduction targets. The changes in the Dutch law are intended to facilitate new developments that will help accelerate the energy transition. However, the energy system for which we are developing legislation and regulation stops at the country's borders, while in the physical reality our national energy system is interconnected with the European gas and electricity networks. Having a comprehensive approach here, would help to connect and align the new policies and regulation with European and international new policies; and this will accelerate the energy transition as well as result in a more coherent energy system—both physically and institutionally—worldwide. Our scientific research of *Complex Socio-Technical Systems Engineering* contributes significantly to this comprehensive approach as it transcends the scale levels of districts, cities, countries and continents to consider other boundaries as well.

Finally, we would like to draw attention to one of the 17 Sustainable Development Goals (SDGs): ensure access to affordable, reliable, sustainable and modern energy for all. The European Pillar of Social Rights, which was established by the European Parliament at the end of 2017, declares that all European citizens are entitled to reliable and affordable energy services (The European Pillar of Social Rights, 2020). We take this for granted in the developed countries, but it is by no means a given fact in many countries in the world. Researchers of the University of Oxford have shown that (Ritchie, 2019):

- nearly 13% of the world's population still have no access to electricity
- energy consumption is not evenly distributed and varies more than tenfold across the world, and electricity consumption more than 100-fold
- energy access is strongly related to income.

To what extent do we need to accept responsibility for this inequality? We will leave this question open; it is the question we ask ourselves frequently when looking for smarter and more effective solutions for the energy transition in the Dutch and European context.

In the next section, we discuss one of potentially effective solutions for the energy transition and for enabling access to reliable and affordable energy services, based on using hydrogen as a main energy carrier in our energy system. *Complex Socio-Technical Systems Engineering* teaches us that the rapid emergence of the hydrogen economy requires concerted action amongst the stakeholders along the hydrogen value chain, shifting from local to regional to international perspective. Large-scale deployment of renewable power needs to be followed by large-scale conversion, transport and storage from source to use. From a western European viewpoint, this involves both local production and remote import through gas pipe infrastructures and through shipping. The end users need to invest in ways of adopting large-scale hydrogen for energy and feedstock sources in a cross-sectoral context of industry, mobility and the built environment. The roles of the many actors involved in the value chain should be investigated to understand how individual decision making of the actors can be influenced to result in rigorous actions towards a trans-national, high-impact, adaptive investment and policy agenda to develop an inclusive, fair, reliable, affordable and sustainable hydrogen economy.

Hydrogen's Role in the Energy Transition

As van Wijk emphasises in the first chapter of Part II, hydrogen can play an important role in the global inclusive energy transition. Hydrogen offers both flexibility and reliability of energy supply when being used as a means of storing solar and wind energy, and as a means of transporting energy to the demand location, anywhere in the world, by using the existing gas networks (after some adjustments have been made), by constructing new hydrogen networks, or by ships and trailers.

In this vision of the future energy system, green hydrogen produced using renewable energy sources could well prove to be an essential building block of this transition. To achieve the Paris climate targets, we definitely need to investigate to what extent priority should be given to hydrogen applications (Hydrogen Council, 2020).

To achieve energy transition aimed at decarbonising energy demand and energy production, a complementary energy system based on hydrogen can be developed to provide the necessary flexibility by:

- enabling cost-effective energy storage both in the short- and long-term;
- allowing energy and, more specifically, electricity consumption and production to be decoupled both geographically and temporally;
- helping to stabilise energy prices in the face of variable wind and solar power production;
- facilitating cost-efficient bulk transport over long distances using pipelines, ships and trucks.
- making energy sector and regional coupling possible.

Next to these roles, hydrogen can be used to decarbonise hard to abate energy use in the transport sector, industrial energy consumption, feedstock for industry and heat and power generation in the built environment. The Hydrogen Council has defined seven roles that hydrogen can play in the energy transition (Hydrogen Council, 2020) (Fig. 5).

As presented in the European Green Deal in December 2019 (European Green Deal, 2020) and later in the Hydrogen Strategy for Europe in July 2020 (European Commission, 2020), there is more research and innovation on clean hydrogen needed to tackle climate and environment-related challenges.

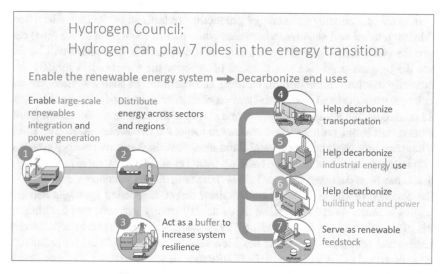

Fig. 5 Roles of hydrogen in the energy transition (Hydrogen Council, 2020)

Car as Power Plant—An Integrated Energy and Mobility System

Many research groups worldwide have been actively researching how to efficiently deploy hydrogen to make the energy, industry, built environment and transport sectors more sustainable. In the transition towards a low carbon energy system, next to finding new energy carriers and sources as well as improving the energy efficiency in different parts of the energy system, *sector coupling* is one of the important aspects. Sector coupling refers to joining efforts between different sectors to support each other and improve overall efficiency and services. In this context, hydrogen can be used to strengthen the integration between energy (electricity, gas, heat) and transport sectors as well as to be deployed efficiently. It means that the system boundary is extended here to include the energy and mobility system with all their domains and interactions.

An example of such an integrated system is shown in the *Car as Power Plant (CaPP)* project, in which hydrogen fuel cell vehicles (FCEVs) are used in the vehicle-to-grid mode to support the energy system (Farahani, 2019). In such a system, the FCEVs do not only provide clean transportation; they can also feed-back electricity to the grid at the times the vehicles are parked. As such, the energy flexibility of hydrogen cars can be cleverly deployed to decouple electricity production and consumption. In short, the CaPP project investigates the potential impacts and feasibility of an integrated energy and transport system consisting of a power system based on wind and solar power, conversion of renewable energy (surpluses) to hydrogen using electrolysis, hydrogen storage and distribution, and hydrogen fuel cell vehicles that provide mobility, electricity, heat and water.

In terms of technology, the energy production system can be envisaged as a fleet of hydrogen fuel cell vehicles, where cars while parked (over 90% of the time) can produce (with their fuel cells) electricity, heat and fresh water, which will be fed into the respective grids. From a social perspective the stakeholders directly and indirectly involved in the design, building and operation of such a system, are car park operators, the local power, heat and water distribution companies, gas suppliers, H2 producers, the equipment, system and software manufacturers but also municipalities, regulators, policy makers and not to forget the car owners/users. The CaPP system has been designed for several stand-alone, distributed, smart energy systems, such as in microgrids or an office building (Alavi et al., 2017; Farahani et al., 2020); also, it has been designed for smart cities (Farahani, 2019; Oldenbroek et al., 2017). The obtained results in different studies show that storage using hydrogen and salt caverns is much cheaper than using large battery storage systems; that the integration of electric vehicles into the electricity network is technically and economically feasible and that they can provide a flexible energy buffer; and that V2G is a promising technology and FCEVs give more flexibility than standard battery EVs since beside storage, they can operate as dispatchable power plants independent of the electricity grid. Ultimately, the results of these studies show that using both electricity and hydrogen as energy carriers can create a more flexible, reliable and cheaper energy system.

The CaPP system with the proposed multi-modelling framework is an example of a carbon-free energy system offering sector coupling and facilitating the penetration of 100% intermittent renewables without any compromise on reliability of energy supply for power, heat and transport and at the same time reducing system cost. Moreover, this approach will engage consumers to have a more active role in the energy transition as prosumers. However, realizing the CaPP concept cannot be done overnight. It requires combining different disciplines to provide comprehensive models and tools supported by real-life pilot projects. We need to provide a single comprehensive framework from different perspectives, such as technical, economic, operational, and social aspects, for designing such a complex socio-technical system (Alavi et al., 2017; Farahani, 2019; Farahani et al., 2020; Oldenbroek et al., 2017; Park Lee, 2019). The emphasis is on the fact that the system design and operation are deeply intertwined and that a stand-alone technical, economic, and social analysis is incomplete without the other ones. Furthermore, new policies to be defined for a carbon-free energy transition are manifold and policymakers require broader knowledge from different disciplines to address the challenges of such a system transition. The CaPP framework stresses the need to consider different aspects such as technology, economics, control, institutional and social perspectives in modelling energy systems. As such, it provides a clearer and more comprehensive insight into the realization of sustainable energy systems to policymakers, compared to the individual models.

Figure 6 shows the physical connection of a hydrogen fuel cell vehicle to the local electricity grid at the lab facility of The Green Village at Delft University of Technology, as part of the demostration phase of the CaPP project.

Fig. 6 The Europe's first Vehicle-to-grid connection using hydrogen fuel cell electric vehicle and the local electricity grid at The Green Village, Delft University of Technology

This innovative *Car as Power Plant* system can make an important contribution to the energy transition. However, to scale the research up to the level of the global energy transition, where hydrogen plays a prominent role, more research is needed. This does not only encompass technological innovation, but also:

- new hydrogen supply chains, from production to the end user
- new infrastructures
- new markets
- new legal and regulatory institutions, including new forms of energy contracts and incentives.

All these changes will not only need to help us to achieve our environmental targets, they will also need to do justice to social expectations and values. To make the realisation of such integrated complex systems possible, large-scale computational models are required. It is clear that these cannot be developed by a single research group alone. The modelling world needs to make a transition as well. We need to develop system models that do justice to many perspectives and many shades of complexity, such as:

- the social, economic and technical complexity of society's energy and mobility systems
- the geographical distribution of these systems
- the dynamic and adaptive behaviour of these systems.

We can meet this need by deploying a multi-modelling framework, whereby various models are developed by various research groups at different locations, and sometimes in different parts of the world, and combined to achieve a common goal. Now is the time for a paradigm change in modelling: A Multi-Modelling Framework for the energy transition, as is called for in (Bollinger et al. 2018).

Conclusion

In this chapter, we have discussed the importance and necessity of a Complex Socio-Technical Systems Engineering approach to achieve an inclusive energy transition. A large network of actors is involved in the development and operation of the future carbon-free energy system, from its technical infrastructure and physical components to new institutions and regulations. New policies to be defined for carbon-free energy transition are manifold and policymakers require broader knowledge from different disciplines to address the challenges of such a radical system transition. The Complex Socio-Technical Systems Engineering framework stresses the need to consider different aspects such as technology, economics, cyber-physical, institutional and social perspectives in modelling energy systems. As such, it provides more comprehensive insight into the realization of new energy systems and richer evidence to inform policymakers, compared to the individual models. Moreover, to realize a low-carbon energy system, different sectors, such as energy (i.e., electricity, heat, gas) and mobility, must support each other to provide reliable services. This so-called sector coupling adds a new layer of complexity which can be tackled by the Complex Socio-Technical Systems Engineering framework.

To illustrate this framework, we have explained how the Car as Power Plant (CaPP) system works. This system is designed as a 100% renewable integrated energy and transport system based on wind and solar power, hydrogen and fuel cell electric vehicles (FCEVs). In the CaPP project, by using techno-economic analysis, we have shown that such a design is technically feasible. However, technical feasibility cannot be guaranteed without considering the controllability of the system. So, the next challenge was to maintain the supply–demand balance as well as to minimize the operational costs of the energy system, which we accomplished by using advanced control techniques. We stress that operation of the innovative CaPP concept should be accompanied by an institutional analysis and designing an organizational system structure. To this end, we have studied the system behaviour and analysed the interactions between different actors in such a system.

The CaPP system and our combined framework is an example of how a comprehensive engineering approach can be used to design an energy system offering sector coupling and facilitating the penetration of 100% intermittent renewables without any compromise on the reliability of energy supply for power, heat and transport, while at the same time reducing system cost. Moreover, this approach will engage energy consumers to have a more active role in the energy transition as well. Hence, by offering our students a program in Complex Socio-Technical System Engineering, not only we will make them smarter and more inclusive engineers and scientists, but also, we can show them the way to achieving an inclusive energy transition.

To conclude, we would like to cite the words of the famous economist John Maynard Keynes (John, 1842). Almost 100 years ago, he described the qualities of a good economist. We are convinced that this description also applies to the inclusive comprehensive engineers and scientists of today:

> He must be mathematician, historian, statesman, philosopher—in some degree. He must understand symbols and speak in words… He must study the present in the light of the past

for the purposes of the future. No part of man's nature or his institutions must be entirely outside his regard.

References

Alavi, F., Park Lee, E., van de Wouw, N., De Schutter, B., & Lukszo, Z. (2017). Fuel cell cars in a microgrid for synergies between hydrogen and electricity networks. *Applied Energy, 1.*

Amer Power Plant, Retrieved on February 2020, from https://www.group.rwe/en/our-portfolio/our-sites/amer-power-plant

Bollinger, L. A., Davis, C. B., Evins, R., Chappin, E. J. L., & Nikolic, I. (2018). Multi-model ecologies for shaping future energy systems: Design patterns and development paths. *Renewable and Sustainable Energy Reviews, 82*, Part 3.

Decision of the Dutch Government: Closure of Hemweg-8 power plant by the end of 2019. Retrieved on February 2020, from https://group.vattenfall.com/press-and-media/news--press-rel eases/pressreleases/2019/decision-of-the-dutch-government-closure-of-hemweg-8-power-plant-by-the-end-of-2019

Electricity Act Amendment (progress of energy transition). (2018) (in Dutch) https://wetten.ove rheid.nl/BWBR0040852/2020-01-01

European Commission. (2020, July). A hydrogen strategy for a climate neutral Europe.

European Green Deal. Retrieved on February 2020, from https://ec.europa.eu/info/publications/ communication-european-green-deal_en

Farahani, S. S., van der Veen, R., Oldenbroek, V., Alavi, F., Park Lee, E. H., van de Wouw, N., van Wijk, A., De Schutter, B., & Lukszo, Z. (2019). A hydrogen-based integrated energy and transport system. *IEEE Systems, Man, and Cybernetics Magazine, 5*(1).

Farahani, S. S., Bleeker, C., van Wijk, A., & Lukszo, Z. (February 2020). Integrated energy and mobility system for a real-life office environment. *Journal of Applied Energy, 264*, 1–21

Hydrogen scaling up. A sustainable pathway for the global energy transition. Hydrogen Council. Retrieved on February 2020, from https://hydrogencouncil.com/en/study-hydrogen-scaling-up/

IRENA. (2018). *Global energy transformation—A roadmap to 2050.* International Renewable Energy Agency.

Keynes, J. M. (1924). Alfred Marshall, 1842–1924. *The Economic Journal 34*(135).

Mittal, S., Ruth, M., Pratt, A., Lunacek, M., Krishnamurthy, D., & Jones, W. *A system-of-systems approach for integrated energy systems modeling and simulation.* Retrieved on February 2020, from https://www.nrel.gov/docs/fy15osti/64045.pdf

Moncada, J. A., Lee, E. H. P., Guerrero, G. D. C. N., Okur, O., Chakraborty, S. T., & Lukszo, Z. (2017). Complex systems engineering: Designing in sociotechnical systems for the energy transition. *EAI Endorsed Transactions on Energy Web, 17*(11), art. no. e1. https://doi.org/10. 4108/eai.11-7-2017.152762.

Oldenbroek, V., Verhoef, L. A., & van Wijk, A. J. M. (2017). Fuel cell electric vehicle as a power plant: Fully renewable integrated transport and energy system design and analysis for smart city areas. *International Journal of Hydrogen Energy, 42*(12), 8166–8196

Park Lee, E. (2019). A socio-technical exploration of the car as power plant. Ph.D. Thesis, TU Delft.

Ritchie, H. (2019). Access to energy, Published online at OurWorldInData.org. Retrieved from: https://ourworldindata.org/energy-access

Senge, P. (1990). *The fifth discipline.* Doubleday.

Sosa, M. E., Eppinger, S. D., & Rowles, C. M. (2003). Identifying modular and integrative systems and their impact on design team interactions. *ASME Journal of Mechanical Design, 125*, 240–252

The European Pillar of Social Rights. Retrieved on February 2020, from https://ec.europa.eu/com mission/publications/european-pillar-social-rights-booklet_en

Printed in the United States
by Baker & Taylor Publisher Services